Associations and Correlations

Unearth the powerful insights buried in your data

Lee Baker

Associations and Correlations

Author: Lee Baker

Managing Editor: Rutuja Yerunkar

Acquisitions Editor: Bridget Neale

Production Editor: Shantanu Zagade

Editorial Board: David Barnes, Mayank Bhardwaj, Ewan Buckingham, Simon Cox, Mahesh Dhyani, Taabish Khan, Manasa Kumar, Alex Mazonowicz, Douglas Paterson, Dominic Pereira, Shiny Poojary, Erol Staveley, Ankita Thakur, and Jonathan Wray

First Published: June 2019

Production Reference: 1280619

ISBN: 978-1-83898-041-2

Published by Packt Publishing Ltd.

Livery Place, 35 Livery Street

Birmingham B3 2PB, UK

Table of Contents

Preface

About

This section briefly introduces the author and coverage of the book.

About the Book

Associations and correlations are perhaps the most used of all statistical techniques. Consequently, they are possibly also the most misused.

The problem is that the majority of people that work with association and correlation tests are not statisticians and have little or no statistical training. That's not a criticism, but simply an acknowledgement that most researchers, scientists, healthcare practitioners, and other investigators are specialists in things other than statistics and have limited – if any – access to statistical professionals for assistance.

Therefore, they turn to statistical textbooks and perceived knowledge among their peers for their training. I won't dwell too much on perceived knowledge, other than to say that both the use and misuse of statistics passes through the generations of researchers equally. There is a lot of statistical misinformation out there...

There are many statistical textbooks that explain everything you need to know about associations and correlations, but here is the rub: most of them are written by statisticians that understand, in great depth, how statistics work and they don't understand why non-statisticians have difficulty with stats – they have little empathy. Consequently, many of these textbooks are full of highly complex equations explaining the mathematical basis behind the statistical tests, are written with complicated statistical language that is difficult for the beginner to penetrate, and they don't take into account that the reader just might be looking into statistics for the first time.

Ultimately, most statistics books are written *by* statisticians *for* statisticians.

In writing this book, I was determined that it would be different.

This is a book for beginners. My hope is that more experienced practitioners might also find value in it, but my primary focus here is on introducing the essential elements of association and correlation analyses. If you want the finer points, then you're plum out of luck – you won't find them here. Just the essential stuff. For beginners.

There's another issue I've noticed with most statistical textbooks, and I'll use a house building analogy to illustrate it.

When house builders write books about how to build houses, they don't write about hammers and screwdrivers. They write about how to prepare the foundations, raise the walls, and fit the roof. When statisticians do their analyses, they think like the house builder. They think about how to pre-process their data (prepare the foundations), do preliminary investigations to get a 'feel' for the data (raise the walls and see what the whole thing will look like), and, finally, they deduce the story of the data (making the build watertight by adding a roof).

Unfortunately, that's not how they write books. Most statistical textbooks deal with statistical tests in isolation, one by one. They deal with the statistical tools, not the bigger picture. They don't tend to discuss how information flows through the data, nor how to create strategies to extract the information that tells the story of the whole dataset.

Here, I discuss a holistic method of discovering the story of all the relationships in your data by introducing and using a variety of the most used association and correlation tests (and helping you to choose them correctly). The holistic method is about selecting the most appropriate univariate and multivariate tests and using them together in a single strategic framework to give you confidence that the story you discover is likely to be the true story of your data.

The focus here is on the utility of the tests and on how these tests fit into the holistic strategy. I don't use any complicated math (OK, well, just a little bit toward the end, but it's necessary, honest...), I shed light on complicated statistical language (but I don't actually use it – I use simple, easy-to-understand terminology instead), and I don't discuss the more complex techniques that you'll find in more technical textbooks.

Chapter 1, Data Collection and Cleaning, briefly (*very* briefly) introduces data collection and cleaning, and outlines the basic features of a dataset that is fit for purpose and ready for analysis.

Chapter 2, Data Classification, discusses how to classify your data and introduces the four distinct types of data that you'll likely have in your dataset.

Chapter 3, Introduction to Associations and Correlations, introduces associations and correlations, explains what they are, and their importance in understanding the world around us.

Chapter 4, Univariate Statistics, discusses the univariate statistical tests that are common in association and correlation analysis, and details how and when to use them, with simple easy-to-understand examples.

Chapter 5, Multivariate Statistics, introduces the different types of multivariate statistics, and how and when to use them. This chapter includes a discussion of confounding, suppressor, and interacting variables, and what to do when your univariate and multivariate results do not concur (spoiler alert: the answer is not panic!).

Chapter 6, Vizualising Your Relationships, explains the holistic strategy of discovering all the independent relationships in your dataset and describes why univariate and multivariate techniques should be used as a tag team. This chapter also introduces you to the techniques of visualizing the story of your data.

Chapter 7, Bonus: Automating Associations and Correlations, is a bonus chapter that explains how you can discover all the associations and correlations in your data automatically, and in minutes rather than months.

I hope you find something of value in this book. Whether you do (or don't) I'd love to hear your comments. Send me an email at: ebookfeedback@chi2innovations.com.

About the Author

Lee Baker is an award-winning software creator that lives behind a keyboard in a darkened room. Illuminated only by the light from his monitor, he aspires to finding the light switch.

With decades of experience in science, statistics, and artificial intelligence, he has a passion for telling stories with data, yet despite explaining it a dozen times, his mother still doesn't understand what he does for a living.

Insisting that data analysis is much simpler than we think it is, he authors friendly, easy-to-understand books that teach the fundamentals of data analysis and statistics – like this one!

As the CEO of Chi-Squared Innovations, one day he'd like to retire to do something simpler, such as crocodile wrestling.

Learning Objectives

- Identify a dataset that's fit for analysis using its basic features
- Understand the importance of associations and correlations
- Use multivariate and univariate statistical tests to confirm relationships
- Classify data as qualitative or quantitative and then into the four subtypes
- Build a visual representation of all the relationships in the dataset
- Automate associations and correlations with CorrelViz

Audience

This is a book for beginners – if you're a novice data analyst or data scientist, then this is a great place to start. Experienced data analysts might also find value in this title, as it will recap the basics and strengthen your understanding of key concepts. This book focuses on introducing the essential elements of association and correlation analysis.

Approach

This book is written with a focus on understanding data, choosing the right ways to analyze it, selecting the correct statistical tools, and interpreting the results in an easy-to-understand way. This book enables you to understand and critically evaluate the results of analyses, and is packed with visually intuitive examples and realistic use cases, making it perfect for beginners.

Data Collection and Cleaning

The first step in any data analysis project is to collect and clean your data. If you're fortunate enough to have been given a perfectly clean dataset, then congratulations – you're well on your way. For the rest of us, though, there's quite a bit of grunt work to be done before you can get to the joy of analysis (yeah, I know, I really must get a life...).

In this chapter, you'll learn about what the features of a good dataset look like and how the dataset should be formatted to make it amenable to analysis by association and correlation tests.

Most importantly, you'll learn why it's not necessarily a good idea to collect sales data on ice cream and haemorrhoid cream in the same dataset.

If you're happy with your dataset and quite sure that it doesn't need cleaning, then you can safely skip this chapter. I won't take it personally – honest!

Data Collection

The first question you should be asking before starting any project is "What is my question?" If you don't know your question, then you won't know how to get an answer. In science and statistics, this is called having a hypothesis. Typical hypotheses might be:

- Is smoking related to lung cancer?

- Is there an association between sales of ice cream and haemorrhoid cream?

- Is there a correlation between coffee consumption and insomnia?

It's important to start with a question, because this will help you decide what data you should collect (and what data you shouldn't).

It's not usual that you can answer these types of question by collecting data on just those variables. It's much more likely that there will be other factors that may have an influence on the answer and all of these factors must be taken into account. If you want to answer the question is smoking related to lung cancer? then you'll typically also collect data on age, height, weight, family history, genetic factors, and environmental factors, and your dataset will start to become quite large in comparison with your hypothesis.

So, what data should you collect? Well, that depends on your hypothesis, the perceived wisdom of current thinking, and any previous research carried out, but ultimately, if you collect data sensibly, you will likely get sensible results and *vice versa*, so it's a good idea to take some time to think it through carefully before you start.

I'm not going to go into the finer points of data collection and cleaning here, but it's important that your dataset conforms to a few simple standards before you can start analyzing it.

By the way, if you want a copy of my book *Practical Data Cleaning*, you can get a free copy of it by following the instructions in the tiny little advert for it at the end of this section...

Dataset Checklist

OK, so here we go. Here are the essential features of a ready-to-go dataset for association and correlation analysis.

Your dataset is a rectangular matrix of data. If your data is spread across different spreadsheets or tables, then it's not a dataset, it's a database, and it's not ready for analysis:

- Each column of data is a single variable corresponding to a single piece of information (such as age, height, or weight, in this case).

- Column 1 is a list of unique consecutive numbers starting from one. This allows you to uniquely identify any given row and recover the original order of your dataset with a single sort command.

- Row 1 contains the names of the variables. If you use rows 2, 3, 4, and so on as the variable names, you won't be able to enter your dataset into a statistics program.

- Each row contains the details for a single sample (patient, case, test tube, and so on).

- Each cell should contain a single piece of information. If you have entered more than one piece of information in a cell (such as date of birth and their age), then you should separate the column into two or more columns (one for date of birth, another for age).

- Don't enter the number zero into a cell unless what has been measured, counted, or calculated results in the answer zero. Don't use the number zero as a code to signify "No Data". By now, you should have a well-formed dataset that is stored in a single Excel worksheet. Each column should be a single variable, with row 1 containing the names of the variables, and below this, each row should be a distinct sample or patient. It should look something like *Figure 1.1*.

UniqueID	Gender	Age	Weight	Height	...
1	Male	31	76.3	1.80	...
2	Female	57	59.1	1.91	...
3	Male	43	66.2	1.86	...
4	Male	24	64.1	1.62	...
...	...	...	...	...	...

Figure 1.1: A typical dataset used in association and correlation analysis

For the rest of this book, this is how I assume your dataset is laid out, so I might use the terms variable and column interchangeably, the same going for the terms row, sample, and patient.

Data Cleaning

Your next step is cleaning the data. You may well have made some entry errors and some of your data may not be useable. You need to find such instances and correct them. The alternative is that your data may not be fit for purpose and may mislead you in your pursuit of the answers to your questions.

Even after you've corrected the obvious entry errors, there may be other types of errors in your data that are harder to find.

Check That Your Data Is Sensible

Just because your dataset is clean, it doesn't mean that it is correct – real life follows rules, and your data must follow them, too. There are limits on the heights of participants in your study, so check that all data fits within reasonable limits. Calculate the minimum, maximum, and mean values of variables to see whether all values are sensible.

Sometimes, putting together two or more pieces of data can reveal errors that can otherwise be difficult to detect. Does the difference between date of birth and date of diagnosis give you a negative number? Is your patient over 300 years old?

Figure 1.2 gives you a list of the most useful measures that will help you discover errors in your data and find out whether real-life rules have been followed.

Numerical Data	Categorical Data
Count of all entries Maximum value Minimum value Number of positive values Number of negative values Number of zeros Number of empty cells Difference between dates	Count of all entries Count of each category Number of empty cells

Figure 1.2: Essential descriptive statistics

Check That Your Variables Are Sensible

Once you have a perfectly clean dataset it is relatively easy to compare variables with each other to find out whether there is a relationship between them (the subject of this book). But just because you can, it doesn't mean that you should. If there is no good reason why there should be a relationship between sales of ice cream and haemorrhoid cream, then you should consider expelling one of or both of those variables from the dataset. If you've collected your own data from original sources, then you'll have considered beforehand what data is sensible to collect (you have, haven't you?), but if your dataset is a pastiche of two or more datasets, then you might find strange combinations of variables.

You should check your variables before doing any analyses and consider whether it is sensible to make these comparisons.

So, now you have collected your data, cleaned your data, and checked that your data is sensible and fit for purpose. In the next chapter, we'll go through the basics of data classification and introduce the four types of data.

Data Classification

Data Classification

In this chapter, you'll learn the difference between quantitative and qualitative data. You'll also learn about ratio, interval, ordinal, and nominal data types, and what operations you can perform on each of them.

Quantitative and Qualitative Data

All data is either quantitative – measured with some kind of measuring implement, such as a ruler, jug, weighing scales, stopwatch, thermometer, and so on – or is qualitative: an observed feature of interest that is placed into categories, as in health (healthy, sick), and opinion (agree, neutral, disagree).

Quantitative and qualitative data can be sub-divided into four further classes of data – Ratio, Interval, Ordinal, and Nominal – as shown in *Figure 2.1.*

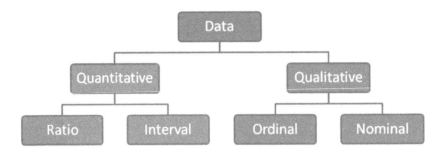

Figure 2.1: There are four distinct types of data

The differences between them can be established by asking just three questions:

Question 1: Are adjacent data points or categories **ordered**?

All measured data is ordered, but not all categories are. If your categories are named [Small; Medium; Large], then there is an order to them. If you have named your categories [1; 2; 3], then there *may* be an order, but it all depends on what 1, 2, and 3 signify. Just because you've used *numbers* to *name* your categories, it doesn't necessarily follow that there is an order.

Question 2: Are adjacent data points **equidistant**?

Look at a ruler and you'll see that the distance between each centimeter marker is precisely the same irrespective of which part of the ruler you're looking at – every centimeter measurement has the same length. In fact, all measuring implements – such as rulers, stopwatches, and jugs – have equidistant data points. Categories, though, do not. The difference in sizes between small and medium is not necessarily the same as that between medium and large. What do you mean by size? Width, height, or depth? Area or volume? It is likely that the reason your data is organized into categories is that it is difficult to accurately measure the feature of interest; therefore, categorical data does not have equidistant data points.

Question 3: Does the scale of measurement have a **meaningful zero**?

All categorical data have arbitrarily chosen zero points. Extrapolate backward from the categories small, medium, and large – where does the line cross the x-axis? Well, it doesn't – it's silly! Similarly, not all measured data have meaningful zero points either. It's easy to see that stopwatches, rulers, and jugs all have meaningful zeros, because you can see them. On the other hand, you can see the zero on a thermometer, too, but the measurement of 0°C is based on the melting point of water at a specific measurement of barometric pressure. Change the pressure, and you change the melting point of water – in other words, 0°C is not fixed!

Figure 2.2 will help you to decide to which data type your data belongs.

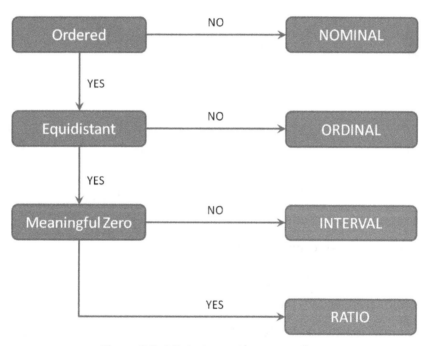

Figure 2.2: 4 Data types, three questions

Nominal Data

Nominal data is observed, not measured, is unordered, is non-equidistant, and has no meaningful zero. We can differentiate between categories based only on their names, hence the title "nominal" (from the Latin nomen, meaning "name"). *Figure* 2.3 may help you decide whether your data is nominal.

Figure 2.3: Features of nominal data

Examples of nominal data include:

- Nationality (British, American, Spanish,...)
- Genre/Style (Rock, Hip-Hop, Jazz, Classical,...)
- Favorite color (red, green, blue,...)
- Favorite animal (aardvark, koala, sloth,...)
- Favorite spelling of "favourite" (favourite, favorite)

The only mathematical or logical operations we can perform on nominal data is to say that an observation is (or is not) the same as another (equality or inequality), and we can determine the most common item by finding the mode (do you remember this from high school classes?).

Other ways of finding the middle of the class, such as median or mean, make no sense because ranking is meaningless for nominal data.

If the categories are descriptive (nominal), such as "Red," "Green," or "Blue," it can be useful to separate each category into its own column, such as Red [Yes; No], Green [Yes; No], and Blue [Yes; No]. These are called "dummy" variables and they can be very useful analytically. More on that later...

Ordinal Data

Ordinal data is observed, not measured, is ordered, but is non-equidistant and has no meaningful zero. Ordinal data categories can be ordered (first, second, third, and so on – hence the name "ordinal"), but there is no consistency in the relative distances between adjacent categories. *Figure* 2.4 shows the features of ordinal data.

Figure 2.4: Features of ordinal data

Examples of ordinal data include:

- Socioeconomic status (lower class, middle class, upper class)
- Opinion (agree, mostly agree, neutral, mostly disagree, disagree)
- Tumor grade (1, 2, 3)
- Political orientation (left, center, right)
- Time of day (morning, noon, night)

Mathematically, we can make simple comparisons between the categories, such as more (or less) healthy/severe, and agree more or less, and since there is an order to the data, we can rank the data and compute the median (or mode, but not the mean) to find the central value.

It is interesting to note that in practice some ordinal data is treated as interval data – tumor grade is a classic example in healthcare – because the statistical tests that can be used on interval data (the data meets the requirement of equal intervals) are much more powerful than those used on ordinal data. This is OK as long as your data collection methods ensure that the equidistant rule isn't bent too much.

Interval Data

Interval data is measured and ordered with equidistant items, but has no meaningful zero. Interval data can be continuous (have an infinite number of steps) or discrete (organized into categories), and the degree of difference between items is meaningful (their intervals are equal), but not their ratio. *Figure* 2.5 will help in identifying interval data.

Figure 2.5: Features of interval data

Examples of interval data include:

- Temperature (°C or F, but not Kelvin)
- Dates (1066, 1492, 1776, and so on)
- Time interval on a 12-hour clock (6am, 6pm)

Although interval data can appear very similar to ratio data, the difference is in their defined zero points. If the zero point of the scale has been chosen arbitrarily (such as the melting point of water or from an arbitrary epoch such as AD), then the data cannot be on the ratio scale and must be interval data.

Mathematically, we may compare the degrees of the data (equality/inequality, more/less) and we may add/subtract the values, such as "20°C is 10 degrees hotter than 10°C" or "6pm is 3 hours later than 3pm." However, we cannot multiply or divide the numbers because of the arbitrary zero, so we can't say "20°C is twice as hot as 10°C" or "6pm is twice as late as 3pm".

The central value of interval data is typically the mean (but could be the median or mode), and we can also express the spread or variability of the data using measures such as the range, standard deviation, variance, and/or confidence intervals.

Ratio Data

Ratio data is measured and ordered with equidistant items and a meaningful zero. As with interval data, ratio data can be continuous or discrete, and differs from interval data in that there is a non-arbitrary zero point to the data. The features of ratio data are shown in *Figure* 2.6.

Figure 2.6: Features of ratio data

Examples include:

- Age (from 0 years to 100+)
- Temperature (in Kelvin, but not °C or F)
- Distance (measured with a ruler or another such measuring device)
- Time interval (measured with a stopwatch or similar)

For each of these examples, there is a real, meaningful zero point – for the age of a person (a 12 year old is twice the age of a 6 year old), absolute zero (matter at 200K has twice the energy of matter at 100K), distance measured from a pre-determined point (the distance from Barcelona to Berlin is half the distance of that from Barcelona to Moscow), and time (it takes me twice as long to run the 100m as Usain Bolt but only half the time of my grandad).

Ratio data is the best to deal with mathematically (note that I didn't say *easiest*...) because all possibilities are on the table. We can find the central point of the data by using the mode, median, or mean (arithmetic, geometric, or harmonic) and use all of the most powerful statistical methods to analyze the data. As long as we choose correctly, we can be really confident that we are not being misled by the data and that our interpretations are likely to have merit.

So far, you have collected, cleaned, pre-processed, and classified your data. It's a good start, but there's still a lot more to do. The next three chapters are about building a statistical toolbox that you can use to extract real, meaningful insights from your shiny new dataset.

3

Introduction to Associations and Correlations

In this chapter, you'll start to understand a little about correlations and associations, including what they are and how you use them. You'll also learn the basics of how to build a holistic strategy to discover the story of your data using univariate and multivariate statistics and find out why eating ice cream does not make you more likely to drown when you go swimming.

You'll learn about why statistics can *never* prove that there is a relationship between a pair of variables, via the null hypothesis, p-values, and a sensibly chosen cut-off value, and learn about how to measure the strength of relationships.

What Are Associations and Correlations?

When you can phrase your hypothesis (question or hunch) in the following form, then you are talking about the relationship family of statistical analyses:

- Is smoking *related to* lung cancer?

- Is there an *association* between education level and political orientation?

- Are height and *weight correlated*?

Typically, the terms correlation, association, and relationship are used interchangeably by researchers to mean the same thing. That's absolutely fine, but when you talk to a statistician, you need to listen carefully – when they say correlation, they are most probably talking about a statistical correlation test, such as a Pearson correlation.

There are distinct stats tests for correlations and for associations, but ultimately they are all testing for the likelihood of relationships in the data.

When you are looking for a relationship between two continuous variables, such as height and weight, then the test you use is called a correlation. If one or both of the variables are categorical, such as smoking status (never, rarely, sometimes, often, or very often) or lung cancer status (yes or no), then the test is called an association.

In terms of correlations, if there is a correlation between one variable and another, what that means is that if one of your variables changes, the other is likely to change, too.

For example, say you wanted to find out whether there was a relationship between age and percentage of body fat. You would do a scatter plot of age against body fat percentage and see whether the line of best fit is horizontal. If it is not, then we can say there is a correlation. In this example, there is a positive correlation between age and percentage of body fat; that is, as you get older, you are likely to gain increasing amounts of body fat, as shown in *Figure* 3.1.

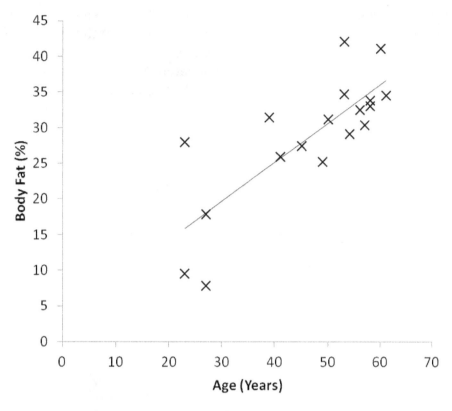

Figure 3.1: Positive correlation with line of best fit

For associations, it is all about the measurement or counts of variables within categories.

Let's have a look at an example from Charles Darwin. In 1876, he studied the growth of corn seedlings. In one group, he had 15 seedlings that were cross-pollinated, and in the other, there were 15 that were self-pollinated. After a fixed period of time, he recorded the final height of each of the plants to the nearest 1/8th of an inch.

To analyze this data, you pool together the heights of all those plants that were cross-pollinated, work out the average height, and compare that with the average height of those that were self-pollinated. A statistical test can tell you whether there is a difference in the heights between the cross-pollinated and self-pollinated groups. This is an example of an association, and is shown as a histogram in *Figure* 3.2.

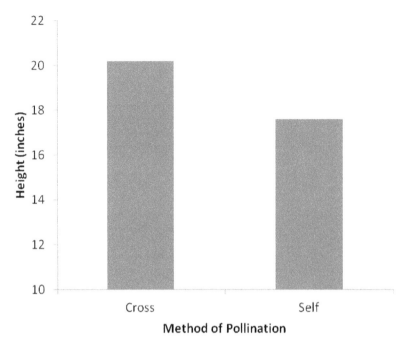

Figure 3.2: Association of heights across two groups

If you don't understand it yet, don't worry – we'll talk more about associations and correlations in the following chapters.

Discovering the Story of Your Data

Let's say you have your hypothesis, you've designed and run your experiment, and you've collected your data.

If you're a medic, an engineer, a businessperson, or a marketer, then here's where you start to get into difficulty. Most people who have to analyze data have little or no training in data analysis or statistics, and round about now is where the panic starts to begin.

What you need to do is:

1. Take a deep breath.
2. Define a strategy of how to get from data to story.
3. Learn the basic tools of how to implement that strategy.

In relationship analyses, the strategy is a fairly straightforward one and goes like this:

- **Clean and classify your data**: You now have a clean dataset ready to analyze.
- **Answer your primary hypothesis**: *This* is related to *that*.
- **Find out what else has an effect on the primary hypothesis**: Rarely is *this* related to *that* without something else having an influence.

Not too difficult so far, is it?

OK, now the slightly harder bit – you need to learn the tools to implement the strategy. This means understanding the statistical tests that you'll use in answering your hypothesis.

Before we go into depth, we need to generalize a little bit.

There are basically two types of test you'll use here:

- Univariate statistics (*Chapter 4, Univariate Statistics*)
- Multivariate statistics (*Chapter 5, Multivariate Statistics*)

They sound quite imposing, but in a nutshell, univariate stats are tests that you use when you are comparing variables one at a time with your hypothesis variable. In other words, you compare *this* with *that* while ignoring all other potential influences.

On the other hand, you use multivariate stats when you want to measure the relative influence of many variables on your hypothesis variable, such as when you simultaneously compare *this*, *that*, and *the other* against your target.

It is important that you use both of these types of test because although univariate stats are easier to use and they give you a good *feel* for your data, they don't take into account the influence of other variables, so you only get a partial – and probably misleading – picture of the story of your data.

On the other hand, multivariate stats *do* take into account the relative influence of other variables, but these tests are much harder to implement and understand and you don't get a good feel for your data.

My advice is to do univariate analyses on your data first to get a good understanding of the underlying patterns of your data, then confirm or deny these patterns with the more powerful multivariate analyses. This way, you get the best of both worlds, and when you discover a new relationship, you can have confidence in it because it has been discovered and confirmed by two different statistical analyses.

When pressed for time, I've often just jumped straight into the multivariate analysis. Whenever I've done this, it always ends up costing me more time – I find that some of the results don't make sense and I have to go back to the beginning and do the univariate analyses before repeating the multivariate analyses.

I advise that you think like the tortoise rather than the hare – slow and methodical wins the race...

Finally, when you have the story of your data, you'll need to explain it to others, whether in person or in a publication of some sort (such as a report, thesis, or white paper).

If your story is a complicated one, then it might be useful to produce some kind of visualization to make it easier for your audience, and some visualizations are more appropriate and easier to understand than others are. We'll explore this in *Chapter 6, Visualizing Your Relationships*.

Correlation versus Causation

And now a word of warning that I'm sure you've heard many times before.

Just because you've found a correlation (or association) between a pair of variables, it does not necessarily mean that one causes the other.

For example, let's say that you discovered in your analysis that there is a relationship between ice cream sales and incidences of drowning. What can you infer from this?

- **Ice cream sales cause people to drown**: Surely not, unless you have to swim a mile out to sea to reach the ice cream vendor's boat.

- **Drowned people eat ice cream in the afterlife**: Well, without a thorough command of how to travel to the afterlife and back again, you will probably never know!

Most likely, there is a variable that is related to both ice cream sales and the likelihood of drowning. It is easy to see in this case that the weather is likely to play a part – more people will eat ice cream and go for a swim to cool off in hot weather than in cold.

Let's not beat around the bush here, though. If you find a relationship between a pair of variables and you've confirmed it with multivariate analysis, then *something* is responsible for the relationship and there *is* a cause. It's your job to find it:

- **It may be that a variable in your dataset is the cause**: Then go and find it – the strategies detailed in this book will help.

- **There might be an influence from something that is not represented in your dataset**: Then you need a new hypothesis (and new data) to find it.

- **The result may have occurred by chance**: Then you need to repeat your experiment to confirm or deny the result.

Assessing Relationships

When you first made your hypothesis, you will have asked yourself the following question:

- Is there a relationship between *this* and *that*?

The answer is absolute – it is either yes or no

This is the question that statistical tests attempt to answer. Unfortunately, they are not equipped to *definitively* answer the question, but rather they tell you how confident you can be about the answer to your question.

The strange thing about statistics is that it cannot tell you whether a relationship is true; it can only tell you how likely it is that the relationship is false. Weird, huh?

What you do is make an assumption that there *is not* a relationship between *this* and *that* (we call this the null hypothesis). Then you test this assumption.

How do we do this?

All relationship tests give you a number between zero and one that you can use to tell you how confident you are in the null hypothesis. This number is called the p-value, and is the probability that the null hypothesis is correct.

If the p-value is large, there is strong evidence that the null hypothesis is correct and we can be confident that there is not a relationship between the variables.

On the other hand, if the p-value is small, there is weak evidence to support the null hypothesis. You can then be confident in *rejecting the null hypothesis* and concluding that there is evidence that a relationship may exist.

OK, so if you have a number between zero and one, how does this help you figure out whether there is a relationship between your variables?

Well, it depends...

Ah, you wanted a definitive answer didn't you? Well, you're not going to get one!

Typically, a p-value of 0.05 is used as the cut-off value between "the null hypothesis is correct" (anything larger than 0.05) and "not correct" (smaller than 0.05).

If your test p-value is 0.15, this corresponds to a probability of 15% that the null hypothesis is correct; we do not reject the null hypothesis (since 0.15 is larger than 0.05) and we say that there is insufficient evidence to conclude that there is a relationship between this and that.

Where did the p-value of 0.05 come from?

This cut-off value has become synonymous with *Ronald A Fisher* (the guy who created Fisher's Exact Test, among other things), who said that:

"...it is convenient to draw the line at about the level at which we can say: "*Either there is something in the treatment, or a coincidence has occurred such as does not occur more than once in twenty trials*"...

As 1 in 20, expressed as a decimal, is 0.05, this is the arbitrarily chosen cut-off point. Rather than being calculated mathematically, it was chosen simply to be sensible.

Since 0.05 is "sensible," then you should also be sensible about test p-values that are close to this value.

You should not definitively reject the null hypothesis when $p = 0.049$ any more than you should definitively accept the null hypothesis when $p = 0.051$.

When your test p-value is close to 0.05, you would be sensible to conclude that such p-values are worthy of additional investigation.

For the purposes of clarity, from this point on, for any p-value smaller than 0.05 we will say things such as "there is (evidence for) a relationship," "is statistically significant," or something similar. Conversely, p-values larger than 0.05 will attract statements such as "there is not a relationship" or "is not statistically significant".

Assessing the Strength of Relationships

OK, so you have tested whether the relationship is likely to be true and concluded – based on the p-value – that it is.

Now you need to test how strongly related the variables are. We call this the effect size, and typical measures of the size of the effect observed are:

- Odds ratio

- Risk ratio

- Hazard ratio

These are all positive values and tell you the size of the difference between two groups.

For example, say that we have a group of 100 men who smoke, and that 27 of them had lung cancer. The odds of getting lung cancer if you're a man are then 27/100 = 0.27.

Now let's look at the group of 213 women who smoke. Say that 57 of those had lung cancer. The odds of getting lung cancer in women are then 57/213 = 0.267.

The ratio of these odds represents the different risks of contracting lung cancer between the gender groups, which is 0.27/0.267 = 1.01.

An effect size close to 1 indicates that there is little or no difference between the groups (and more than likely the p-value would be large and non-significant), that is, the odds are the same in both groups.

What if the odds ratio is 2? This would tell you that male smokers are twice as likely to contract lung cancer as female smokers are.

You can also flip this odds ratio on its head and say that female smokers are half as likely to contract lung cancer as male smokers.

In the next chapter, we'll go through the various kinds of univariate statistics that you'll find in relationship analyses.

Univariate Statistics

In this chapter, you will learn about the different univariate statistical tests that you will use in relationship analysis. You will learn about the different correlation, association, and survival analysis tests in univariate analysis and learn what to look for in a typical set of results to decide whether there is evidence of a relationship between the variables and how strong that relationship is.

Correlations

When both of the variables under investigation are continuous, you ask the question 'is *this* correlated with *that*?'

For example, let's have another look at *Figure* 3.1, where we had a look at the relationship between age and body fat percentage. *Figure* 3.1 is called a scatter plot, and the best-fit line between the plot is called a line of regression.

This is an important graph because it gives you a good feel for your data and what the underlying story might be. When you see a scatter plot such as this, you should always ask the following questions:

- Is there a gradient to the regression line? — Is the gradient positive (slopes up from left to right)?, Or negative (slopes down from left to right)?

- Do the plot points fall close to or far from the regression line?

- Is the linear regression line a good representation of the data? — Would a curve or other line shape be a better fit?

- Are there clearly defined clusters of data?

- Are there any outliers?

Now that you have a good sense of what's going on with the data, you can test whether there is a correlation between age and body fat.

There are two types of correlation test: the Pearson and Spearman correlations. *Figure 4.1* will help you decide which is most appropriate for your data.

PEARSON CORRELATION	SPEARMAN CORRELATION
Assumes that both variables are normally distributed	Makes no assumption about the data distributions
Use this when there is a linear relationship between the variables	Use this when there is a non-linear relationship between the variables
If the assumptions are met, Pearson's correlation is more powerful than Spearman's	When the normality assumptions are not met, Spearman's correlation is more appropriate

Figure 4.1: Choice between the Pearson and Spearman correlation tests

If a straight line is not the best fit for the data, then there are methods that you can try to get a better fit, such as data transformations. In this case, non-linear data may be assessed by the Pearson correlation rather than the Spearman correlation, but these methods are beyond the scope of this book.

Whether you use the Pearson or Spearman correlation test, the important outputs that you will need when you run these tests on your data are:

- **Visualization**: A scatter plot with a best-fit regression line
- **Statistic**: The p-value, correlation coefficient R (or R^2)
- **Effect Size**: The equation of the line (the gradient and intercept)

The scatter plot is important because it tells you whether there is likely to be evidence of a relationship between the pair of variables, and which type of correlation test is likely to be the better choice.

The p-value is important because it tells you whether there is evidence of a relationship between the pair of variables.

If the p-value is not significant, then the correlation coefficient and equation of the line are meaningless – if you have concluded that there is not a relationship between the variables, then measuring the strength of the relationship makes no sense.

The correlation coefficient gives you a measure of how well the plot points fit the regression line.

If the correlation coefficient R is large, then the points fit the line very well.

Let's say that R = 0.964. You can say that 93% ($R^2 = 0.964^2 = 0.93$) of the variation in outcome is explained by the predictor variable. On the other hand, if R = 0.566, then only 32% ($R^2 = 0.566^2 = 0.32$) of the variation is due to the predictor and you need to look at other variables to try to find alternative explanations for the variation in outcome.

The equation of the line, *outcome = (gradient × predictor) + intercept* gives you a way to predict one of the variables if you know the value of the others.

The gradient is useful as it tells you by how much the outcome should increase for a single unit change in the predictor and is a measure of the size of the effect.

OK, let's run a Pearson correlation of the age and body fat percentage data. The results are shown in *Figure 4.2*.

Regression equation:

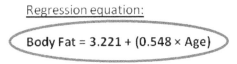

Body Fat = 3.221 + (0.548 × Age)

Summary Table:

Predictor	Coef	SE Coef	T	P
Constant	3.2210	5.0760	0.63	0.535
Age	0.5480	0.1056	5.19	<0.001

Regression Statistics:

S	R²	R² (adj)
5.75361	62.7%	60.4%

Figure 4.2: Typical results of a Pearson correlation

The Age p-value tells you that there is a significant correlation between age and body fat (p < 0.001) and its coefficient tells you that the correlation is positive – but, of course, you knew that already from the scatter plot, didn't you?

The R^2 value tells you that 62.7% of the variation in the body fat percentage is explained by age (the adjusted R^2 value is only useful when assessing multiple predictor variables and can be ignored in univariate analysis).

Finally, the equation of the line tells you that, on average, body fat increases by an additional 0.548% for every single unit increase in age (that is, for every year).

Associations

While correlations concern only continuous data such as height, weight, age, and so on, associations concern categorical data such as gender [male; female], status of some sort [small; medium; large] and even continuous data that has been arranged in categories, such as height [short; average; tall] and age [30s; 40s; 50s; and so on].

There are two types of association that you need to be aware of:

- Categorical *versus* categorical
- Categorical *versus* continuous

To confuse things just a little bit more, there are two flavors of categorical data:

- Ordinal
- Nominal

Recall from *Chapter 2, Your Statistical Toolbox*, that ordinal data is categorical data that has a natural order to the categories, whereas nominal data categories have no natural order.

In this case, it can be useful to separate the categories into separate variables, each having two categories, as in *Figure 4.3*.

Gender
Male
Female
Male
Transgender
...

Male	Female	Transgender
Yes	No	No
No	Yes	No
Yes	No	No
No	No	Yes
...	...	...

Figure 4.3: Separate the categories of nominal variables into separate columns

It is important to remember this at this stage because you will often analyze data that has more than two categories. Many tools assume that your categories are ordinal, so you really do need to organize your data appropriately and use the tools correctly. If you don't, the results you get will be misleading at best and just plain wrong at worst.

For data with just two categories, it doesn't matter whether they are nominal or ordinal – their treatment and analysis is just the same.

The m×n Categorical

When you are comparing two variables that each have multiple categories, you can call this m×n ('m by n') categorical analysis. It's also sometimes known as r×c categorical analysis. If you find this terminology confusing, you can think of this as the analysis of counts within categories.

Let's have a look at a hypothetical example. Say that you're looking into the highest level of education achieved (graduated from university, high school, or neither) for young adults whose parents are classified according to their socioeconomic status (lower, middle, or upper class). You hypothesize that the wealthier the parents are, the higher the level of education achieved by their children will be.

To visualize what this data looks like, we create a 3×3 grid and put each of the patients into the appropriate cell. There are many alternative names for such a grid, including a frequency table, a contingency table, and a confusion matrix, and it looks like *Figure* 4.4.

		Socioeconomic Status (Class)		
		Lower	Middle	Upper
Education Level	None	165	33	2
	High School	457	122	51
	University	298	174	103

Figure 4.4: A 3×3 contingency table

Although it looks like an important table, it's actually difficult to read and understand if there is likely to be a relationship between the two variables.

What you need is a way to calculate the expected value for each cell, then you can figure out by how much your actual values deviate from what you would expect by chance.

You do this by using these equations: *Expected Cell Value = Column Sum × Row Sum/ Table Total* and x^2 = *(observed – expected)* ^{2} */ expected*.

When you calculate *observed – expected*, some of the cell values will be negative. Make a note of which ones, and when you put the results into the table, add the negative sign back into the appropriate cells.

This will give us what are called the Chi-Squared cell values (x^2) and tell you whether, and by how much, the observed values in each cell differ from what would be expected by chance.

Your table of Chi-Squared cell values will now look like *Figure* 4.5.

		Socioeconomic Status (Class)		
		Lower	Middle	Upper
Education Level	None	8.8	-4.1	-18.4
	High School	4.8	-4.4	-5.1
	University	-16.4	11.5	24.0

Figure 4.5: A 3×3 contingency table of Chi-Squared cell values

This table will now help you to get a good feel for your data, and you should always ask the following questions:

- Which cells are positive (have more involved lymph nodes than is expected by chance)?

- Which cells are negative (have fewer involved lymph nodes than is expected by chance)?

- Are there any cell values that are larger than the rest? – This is the critical cell and is the cell that differs most from what we would expect by chance.

- Are these results what you expected to find?

For this data, you can see that the largest Chi-Squared cell value corresponds to the highest socioeconomic class and the highest possible education level achieved and is positive. This tells you that there are more children of upper-class parents achieving a university education than expected by chance. In fact, the adjacent cells suggest that a university education is more likely than expected by chance for the middle and upper classes (cell values of 11.5 and 24.0) but those in the lower socioeconomic class are less likely to achieve a university education (cell value of -16.4).

The largest negative Chi-Squared cell value (-18.4) suggests that it is highly unlikely that children of upper-class parents will not graduate from any level of education.

Now that you have a good sense of what's going on with your data, you can test whether there is an association between these variables, and for this you need the Pearson Chi-Squared Test – and yes, this is the same Pearson that came up with the correlation test.

The more you delve into stats, the more often you will bump into **Karl Pearson** and **Ronald A. Fisher** (more on him soon!) – these guys are gods in the stats world!

Well, as it turns out, you've already done most of the work for the Chi-Squared Test. All you need to do now is add up all the Chi-Squared cell values (ignoring the sign) and you have the Chi-Squared value for the whole table.

This is the number that you get when you run a **Chi-Squared Test** on your observed values, and it tells you whether there is evidence of an association between your pair of variables.

The important outputs that you will need when you run the Chi-Squared Test on your data are:

- **Visualization**: An m×n summary table and an m×n table of Chi-Squared cell values
- **Statistic**: An x^2 value, the number of degrees of freedom, and a p-value

The summary table is important because it gives you a good feel for the distribution of your data. However, it can be difficult to tell whether any of the cells are significantly different from what should be expected by chance.

The Chi-Squared cell value summary table is important because it tells you whether any of the cells are different from what should be expected by chance and allows you to easily identify the critical cell.

The x^2 value is not so important, because on its own it tells you little about whether or not there is evidence of an association between your variables.

The number of degrees of freedom (df), similarly, is not so important, but when you put the x^2 value and df together, you can calculate the p-value.

The p-value is important because it tells you whether there is evidence of an association between the pair of variables.

Figure 4.6 shows the results of running a Chi-Squared Test on the lymph node data.

Chi-Squared Statistics:

x^2	df	P
97.55	4	<0.001

Figure 4.6: Typical results of a Chi-Squared test

The p-value tells you that, indeed, there is a significant association between socioeconomic status and education level achieved. What the statistic doesn't tell you is what is responsible for the association or whether the association is positive or negative. This is why it's important to get your hands dirty with the data.

Note here that you don't calculate an effect size. Although there are effect size calculations that can be done with m×n tables, there is a lot of debate about how useful they are. What is more useful is to identify a portion of the table from the Chi-Squared cell values and isolate that portion so that you get a 2×2 table.

From here, you would then do a standard analysis for 2×2 tables, which, by a remarkable coincidence, is exactly what we're looking at next!

Read on...

The 2×2 Categorical

Arranging the data for two dichotomous variables (that's stats-speak for 'having two parts') is exactly the same as for analyzing m×n data, but this type of analysis has a special group of statistics that are much more powerful and accurate.

OK, let's look at a different example.

Let's say that you're investigating whether there is any association between students having studied mathematics at pre-university level and whether they went on to study physics or biology at university. Your data is shown in *Figure* 4.7.

		Degree Studied	
		Physics	Biology
Mathematics Qualification	No	83	487
	Yes	19	61

Figure 4.7: A 2×2 contingency table

As with the 3×3 table above, you can work out the Chi-Squared cell values for this table and this will help you to identify whether any particular cell is larger than the others, as in *Figure* 4.8.

		Age	
		Under 50	Over 50
Metastases	No	-0.46	0.09
	Yes	3.31	-0.62

Figure 4.8: A 2×2 contingency table of Chi-Squared cell values

The critical Chi-Squared cell value (bottom-left) tells you that students with a prior mathematical qualification are more likely than expected by chance to study physics at university. Note that the top-left to bottom-right diagonal contains only negative Chi-Squared cell values, while the opposite diagonal contains only positive Chi-Squared cell values. This tells you that there is a negative association between these variables. Of course, if you had presented biology before physics in the table, this would then be a positive association.

With the m×n table, you used Pearson's Chi-Squared Test to figure out whether there was a relationship between the variables. While the Chi-Squared Test is still a fine test to use, it is known to be a little inaccurate, particularly when the counts in any of the cells are small. Various smart mathematicians have tackled this problem over the years and come up with a whole variety of other ways to analyze a 2×2 table more accurately.

Although all of the newer methods are a lot more complicated, these days, it doesn't really matter very much – modern computers do all the grunt work for you and it's just as quick to run Fisher's Exact Test (see, I told you we'd run into Fisher again, didn't I?) as a Chi-Squared Test.

So, let's have a look at a few stats that you might use for a 2×2 table.

The Chi-Squared Test is easiest and quickest to compute but is the least accurate choice. A more accurate choice is the Chi-Squared Test with Yates' Correction.

Better still is Fisher's Exact Test, but this has the same drawback as the Chi-Squared Test, so instead we can use Fisher's Exact Test with Lancaster's mid-p correction.

It may be a rather long name, but this is currently thought to be the most accurate test for 2×2 tables. Alternatives do exist (such as Barnard's Exact Test), but they are computationally much more complex, are not more accurate, and I don't intend to go into them any further here.

Figure 4.9 shows a selection of results from each of these tests for a single 2×2 table. Note how the p-values become smaller the further down the table you go. Well, mostly anyway...

Statistical Results:

	P-value		
	Left tail (negative)	Right tail (positive)	2-tail (both)
χ^2 Test	N/A	N/A	0.034
χ^2 Test with Yates' Correction	N/A	N/A	0.051
Fisher's Exact Test (FET)	0.029	0.986	0.047
FET With Lancaster's Mid-P Correction	0.022	0.978	0.040

Effect Size:

Odds Ratio	Lower 95% CI	Upper 95% CI
0.547	0.301	1.002

Figure 4.9: Some typical results of the analysis of a 2×2 contingency table

So, if there are different stats tests that you could use and they all give slightly different results, which one is right, and which one should you use?

Well, that's the thing about stats – it's not about right or wrong; it's about confidence. Statistics is the study of uncertainty, so you'll never get answers that are black or white.

What I can tell you is that the stats community generally agrees that the results lower down in *Figure* 4.9 are accurate more often than the results above, so the further you go down the table, the more confidence you will have in the result.

My particular preference is mid-p correction and it's the one I recommend using, but you may use any of them as long as you understand their limitations.

The important outputs that you will need when you run 2×2 analysis on your data are:

- **Visualization**: A 2×2 summary table and a 2×2 table of Chi-Squared cell values

- **Statistic**: The p-value

- **Effect Size**: An odds ratio

The summary table is important because it gives you a good feel for the distribution of your data. However, it can be difficult to tell whether any of the cells are significantly different from what should be expected by chance.

The Chi-Squared cell value summary table is important because it tells you whether any of the cells are different from what should be expected by chance and allows you to easily identify the critical cell.

The p-value is important because it tells you whether there is evidence of an association between the pair of variables.

The odds ratio is important because it tells you how strongly related the variables are and represents the different odds between the groups.

I'm sure you noticed that for each of Fisher's Exact Tests, there are three different p-values: the left tailed, right tailed, and two-tailed p-values. 'Which one should I use?', I hear you ask.

Well, if you made a hypothesis before you ran the test, then you should select the one that corresponds with your hypothesis. Having a prior hypothesis means that you can legitimately ignore the two-tailed p-value and choose either the left-tail or right-tailed p-value. If you hypothesized that as one variable goes up the other goes down, then you hypothesized a negative association and you should choose the left-tailed p-value.

Alternatively, if you hypothesized a positive association (as one variable goes up, so does the other), then you should use the right-tailed p-value.

If you had no prior hypothesis at all, then I'm afraid that you're stuck with the two-tailed p-value. No cheating now...

You probably also noticed that the odds ratio is less than one. This confirms that there is a negative association between having a mathematics qualification and the degree studied. For negative odds ratios, you can take the reciprocal to give you a number that is more intuitive, so 1/0.547 gives you an odds ratio of 1.828; that is, university physics students were almost twice as likely to have had a prior mathematics qualification as those that studied biology.

Taking the reciprocal of the confidence intervals tells you that you can be 95% confident that the true odds ratio lies in the range [0.998; 3.322]. Since these confidence intervals span the number one, you cannot be certain that your variable will have any effect on the outcome.

Since the lower confidence interval is close to 1, you cannot definitively accept or reject the null hypothesis based solely on the odds ratio (which you shouldn't anyway), but should take a pragmatic approach and take all results into account; all p-values and odds ratios.

The 42×n categorical

Sometimes, you may find a special case when one of your variables has two categories while another has more than two – but they must be ordinal! You could analyze this by treating it as m×n and simply running a Chi-Squared Test.

However, there is a better way!

What you're looking for here is a clear progression in one group compared to the other, and in this case, the Chi-Squared for Trend is a much more accurate statistic than the plain vanilla Pearson Chi-Squared Test because the order of categories is taken into account as well as the cell counts.

You arrange your data in a 2×n contingency table as before, but rather than work out the Chi-Squared cell values, you work out the proportions in each row.

Let's have a look at an example of a new treatment for Bovine Spongiform Encephalopathy – aka 'Mad Cow Disease' – compared with the standard. The six levels of assessment of a cows' condition are tabulated against the standard and new treatments.

In a standard Chi-Squared test, you would get the same result irrespective of the ordering of categories, but with the Chi-Squared for Trend test, if there is a relationship between the new treatment and the patient condition, you would expect a greater improvement with the new treatment over the standard treatment, and this would be reflected in the proportions, *Figure 4.10*.

		Treatment		
		Standard	New	Proportion
Assessment	Large Improvement	4	28	0.875
	Small Improvement	13	10	0.435
	No Change	3	2	0.400
	Small Deterioration	12	5	0.294
	Large Deterioration	6	6	0.500
	Death	14	4	0.222

Figure 4.10: A 2×6 contingency table with proportions

If you plot the proportions on a scatter plot with a regression line, this will help you see whether there is a clear trend in proportions between the groups.

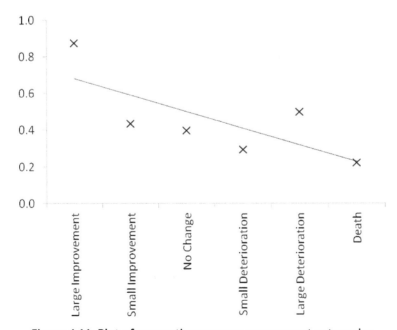

Figure 4.11: Plot of proportions versus assessment categories

At this point, you might be thinking 'Aha, a scatter plot with a regression line – I know how to analyze this; it's the Pearson correlation!'. If you were thinking this, nice try, but get to the back of the class!

The Pearson correlation is for when both of your variables are continuous. Although one of the variables in the plot is now continuous (the proportion), the other is still clearly categorical, so you can't use a correlation test.

We plot the proportion to help us to understand the trend in the data, but both variables are categorical and so the Chi-Squared for Trend is the most appropriate test here.

Figure 4.12 shows the results of running both a Chi-Squared and a Chi-Squared for Trend test on the data in *Figure* 4.10.

Chi-Square Test:

χ^2	df	P
26.97	5	<0.0001

Chi-Square for Trend Test:

χ^2	df	P	Gradient	Intercept
17.93	1	<0.0001	-0.0907	0.7718

Figure 4.12: Chi-Squared results of the analysis of a 2×6 contingency table

In this case, it doesn't really matter whether you ran a Chi-Squared or a Chi-Squared for Trend test, the result is highly significant in either case. You should have noticed though that the proportion in the Large Improvement category is substantially larger than in all other categories, suggesting that the new treatment is significantly more effective than the standard treatment. It would not be inappropriate to have split the table below this category to create a 2×2 contingency table and run the corresponding tests, as in the 2×2 *categorical* section.

To illustrate how important the ordering of categories is in a Chi-Squared for Trend test, try moving the Large Improvement row from the top of *Figure* 4.10 to the bottom, below the Death category, and re-running the analyses.

The Chi-Squared test results will be precisely the same, with a p-value of <0.0001, but the p-value for the Chi-Squared for Trend test will be markedly different at p = 0.255. Just by moving one row, your result goes from highly significant to not-even-close-to-significant.

The important outputs that you will need when you run the Chi-Squared for Trend Test on your data are:

- **Visualization**: A 2×n summary table with proportions and a scatter plot of proportions with a regression line
- **Statistic**: The x^2 for trend value, the p-value
- **Effect Size**: The equation of the line (the gradient and intercept)

The 2×n summary table is important because it gives you a good 'feel' for your data and helps you decide whether there may be a trend.

The scatter plot is important because it gives you a sense of the scale of the trend and whether the plot points are close to or far away from the regression line.

The x^2 value is not so important, because, on its own, it is not always possible to tell whether the observed and expected cell counts are different enough to be statistically significant.

Note that the number of degrees of freedom in a Chi-Squared for Trend Test is always one.

The p-value is important because it tells you whether there is evidence of an association (a trend in the ratio of proportions) between the pair of variables.

The equation of the line, *proportion = (gradient × score) + intercept*, gives us a way to predict one of the variables if we know the value of another.

The n×Categorical versus Continuous

When you have one variable that is categorical while another is continuous, the best way to analyze this is to aggregate all the continuous values into categories. We can think of this as an analysis of measurements by categories, and when you do this, you can compare the differences in the means across the groups.

Let's say that you have collected data on the average number of eggs laid per female fruit fly per day for the first 14 days of life. You had 25 female fruit flies in each of three genetic lines: 25 that were selectively bred for resistance to a specific pesticide, 25 for susceptibility to the same pesticide, and 25 that were non-selected, that is, they were the control group.

So, you will now gather together the average number of eggs laid per female fruit fly per day for the Resistance category. Do this for each of the categories and you'll have three sets of continuous data, representing the Resistant, Susceptible, and Non-Selected categories.

To help you understand how the data is distributed, you can plot box-and-whisker plots of the grouped data, as in *Figure 4.13*.

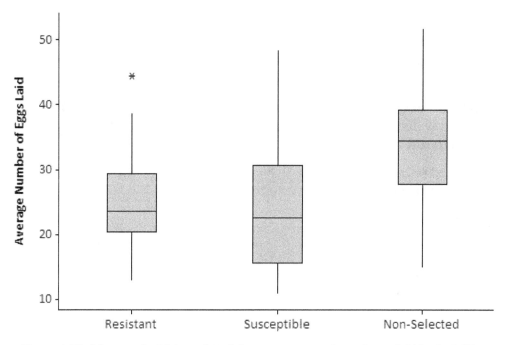

Figure 4.13: A box-and-whisker plot of the average number of eggs laid by fruit flies

The box tells you the first and third quartiles of the data (Q1 and Q3), with the median somewhere between. The limits of the whiskers tell you which data lie within 1.5 inter-quartile ranges (IQR) away from Q1 and Q3:

Lower whisker=Q1-1.5×(IQR) and Upper whisker=Q3+1.5×(IQR) where IQR=Q3-Q1.

Outliers are the asterisks that lie out with the whiskers.

Box-and-whisker plots are great for getting a 'feel' for your data, locating the central point, and deciding whether the data is distributed normally, symmetrically, or skewed.

Looking at the central points and widths of the distributions, they also give you a sense of whether the means are likely to be different (far apart) or similar (close together).

The kinds of questions you should be asking about these plots include:

- Are the individual plots normally distributed? —If they're not normally distributed, are they at least symmetrical?, Are any of the plots skewed?

- Are the means close together or far apart?

- Is there a progression in the means across the groups (if appropriate)? — Are they increasing or decreasing?

- Are the distributions: — Tall and thin (a small variation in values)?, Short and fat (a high variation in values)?

If the distributions are tall and thin, then there is a higher likelihood that the group means are different, even if they are quite close together. On the other hand, if the distributions are short and fat, it's less likely they will be different, even if the means are far apart.

Unless the means are pretty much the same (that is, clearly not different) or they are far apart with narrow distributions (that is, clearly very different), then the only way to find out for sure is to run a statistical analysis.

For this, we need either the ANOVA (**AN**alysis **O**f **VA**riance) or the Kruskall-Wallis Test. *Figure* 4.14 will help you decide which one to choose.

ANOVA	Kruskall-Wallis Test
Assumes that all distributions are normal	Makes no assumption about the data distributions
May still be used if the data are non-normal, but are symmetrically distributed	Use this when the data are non-normal and not symmetrically distributed
If the assumptions are met, the ANOVA is more powerful than the Kruskall-Wallis Test	When the assumptions are not met, the Kruskall-Wallis Test is more appropriate

Figure 4.14: Choice of ANOVA and the Kruskall-Wallis tests

ANOVA, developed by our old friend Ronald A. Fisher, is more accurate than the Kruskall-Wallis test (surprisingly, not developed by Fisher!) but it depends on the data in all groups being normally distributed (or at least symmetrically distributed). If any of the groups fail this test, then confidence in the result will likely be lower and you should use the Kruskall-Wallis test.

If you're unsure which one to use, then the Kruskall-Wallis test will always be valid and you can use it (although, you didn't hear it from me – my membership of GARGS, the Guild of Argumentative Statisticians, may be revoked!).

There is a whole family of different types of ANOVA tests, depending on the number of factors that you're investigating. If you're looking into just one factor, you should use one-way ANOVA. If more than one factor is involved, there are more complicated versions of ANOVA you can use, but these are beyond the scope of this book.

The important outputs that you will need when you run ANOVA or the Kruskall-Wallis test on your data are:

- **Visualization**: A box-and-whisker plot
- **Statistic**: The central values (means or medians) of each group, variation (SD, variance, inter-quartile range or 95% confidence intervals of each group), and the p-value

Note that here we don't calculate an effect size. Although there are effect size calculations that can be done with this type of analysis, there is a lot of debate about how useful they are, as the calculation basically allows you to categorize the effect size as small, medium, or large. Not very useful really...

The box-and-whisker plot is important because it tells you where the central points of the data lie, whether the data is skewed, and which points are outliers.

The central values and measures of variation are important because it is precisely these that are being compared.

The p-value is important because it tells you whether there is evidence of an association between the categories. The other outputs can be useful to understand your data, but only the p-value can truly tell you whether there is evidence of an association.

Figure 4.15 is the output from one-way ANOVA of the fruit fly data.

One-way ANOVA:

	df	SS	MS	F	P
Factor	2	1362.2	681.1	8.67	<0.001
Error	72	5659.0	78.6		
Total	74	7021.2			

Regression Statistics:

S	R^2	R^2 (adj)
8.866	19.40%	17.16%

Descriptive Statistics:

	N	Mean	SD
Resistant	25	25.256	7.772
Susceptible	25	23.628	9.768
Non-Selected	25	33.372	8.942
Pooled SD			8.866

Figure 4.15: Typical results of one-way ANOVA

The p-value tells you that there is a significant difference between the groups and that there is strong evidence that the average number of eggs laid per day is different among the genetic lines.

On its own though, the ANOVA result doesn't tell you which of the groups are different from the others. The box-and-whisker plot can tell you this. It appears from these plots that the resistant and susceptible genetic lines are both significantly different from the non-selected genetic line but are not different from each other. This can be confirmed by analyzing these genetic lines pairwise, which is precisely what we will look at next.

The 2×Categorical versus Continuous

When your categorical variable has two levels, you deal with the data in precisely the same way as if it had multiple levels.

You aggregate all the continuous data into its respective categories and plot histograms and box-and-whisker plots to check for normality, symmetricity, and skewness, and to assess the variability of the data in each group and whether the central values are likely to be different.

Then, you run either the two-sample t-test (if both distributions are normal) or the Mann-Whitney U-test. Figure 4.16 will help you decide which is most appropriate for your data.

2-sample t-test	Mann-Whitney U-test
Assumes that both distributions are normal	Makes no assumption about the data distributions
May still be used if the data are non-normal, but are symmetrically distributed	Use this when the data are both non-normal and not symmetrically distributed
If the assumptions are met, the 2-sample t-test is more powerful than the Mann-Whitney U-test	When the assumptions are not met, the Mann-Whitney U-test is more appropriate

Figure 4.16: Choice of the two-Sample t-test and the Mann-Whitney U-test

FYI – you can run an ANOVA on this data and get exactly the same result as the two-sample t-test. Similarly, the Mann-Whitney U-test will give you the same result as the Kruskall-Wallis test.

The two-sample t-test is also known as the Student's t-test. When **William Sealy Gosset** – employed by Guinness Breweries to improve the quality of stout – was forbidden by his employer from publishing any findings, he instead published his work under the pseudonym 'Student.' And a legend was born…

Anyway, let's have a look at an example. Suppose you wish to find out whether US cars are more economical than Japanese cars, so you measure the number of miles per gallon for a whole bunch of US and Japanese cars under identical conditions. The results of a two-sample t-test are shown in *Figure 4.17*.

Descriptive Statistics:

	N	Mean	SD	SE Mean
US	249	20.14	6.41	0.41
Japanese	79	30.48	6.11	0.69

Regression Statistics:

T	df	P
-12.95	136	<0.001

Figure 4.17: Results of a two-sample t-test

The results here tell you that there is a significant difference (p < 0.001) between US and Japanese cars, and that, in this dataset, Japanese cars run on average 50% further than US cars per gallon.

The important outputs that you will need when you run the two-sample t-test or the Mann-Whitney U-test on your data are:

- **Visualization**: A box-and-whisker plot

- **Statistic**: The central values (means or medians) of each group, variation (SD, variance, inter-quartile range or 95% confidence intervals of each group), and the p-value

Again, we don't calculate an effect size with this analysis. Although there are effect size calculations that can be done, such as Cohen's d, they are not really very useful, as the calculation only allows you to categorize the effect size as small, medium, or large.

The box-and-whisker plot is important because it tells you where the central points of your data lie, whether the data is skewed, and which points are outliers.

The central values and measures of variation are important because it is precisely these that are being compared.

The p-value is important because it tells you whether there is evidence of an association between the pair of variables. The other outputs can be useful to understand your data, but only the p-value can truly tell you whether there is evidence of an association between the variables.

Survival Analysis

Survival analysis is a method for analyzing the time to an event rather than the event itself and is used extensively in medical studies to assess survival rates or disease recurrence rates. The method is also gaining increasing usage and acceptance in many other fields, such as in engineering, to assess the failure rates of mechanical systems or parts, and in many other sectors, such as economics, sociology, and sales forecasts.

If you have a pair of categorical variables, Gender [male; female] and Survival [yes; no], you can simply run Fisher's Exact Test to find out whether men are more likely than women to die (as a result of some disease or other).

Alternatively, if you have a third variable, 'Time To Event,' then you can see how the event in question unfolds over time rather than just a final snapshot at the end of the study, and this is the essence of survival analysis.

A dataset for survival analysis typically looks like *Figure 4.18*.

UniqueID	Survival Status	Survival Time	Component
1	Alive	6.3	A
2	Dead	2.1	B
3	Alive	6.2	B
4	Alive	4.1	A
...	...	...	...

Figure 4.18: A typical dataset used in survival analysis

Let's look at the two survival data variables in question.

you have changed the parameter values, copy and paste : This is the event under study; for instance, the survival status (or otherwise) of an electronic component. In this data column, you note whether the component failed during the period of the study (record 'dead' or 'failed') or survived the study period (record 'alive' or 'survived').

This is also known as the 'censor' variable because it is an observation that is only partially known (that is, failure may occur outside the study period). When you record this variable, your statistics program will ask you which of the data levels corresponds to an 'event,' and which to a 'censor.' It is important to get this right, otherwise the analysis will not be correct. Failure is an event (you can pinpoint exactly when it occurred), whereas alive is not and is the censor.

The 'time to event' variable: It is important that a zero time point be defined for each component in the study.

For example, for survival data, you need to figure out the time from an event (say, time from switch on) to the event in question (which, in this case, would be either failure or the end of the study period).

When collecting such data, you don't typically collect durations; you collect times and dates. For each component, you collect their time and date of switch on. If they fail during the study period, you also collect their date of failure and note their status (dead) in the event column. If they survive to the end of the study, then you record this in the event column (alive). You will then have two date columns that you can subtract to get the time from diagnosis to failure or the end of the study.

This time-to-event variable is known as 'right-censored,' since components surviving beyond (that is, *to the right of*) the end of the study period must be censored, as their time of failure is as yet undetermined. It is important to choose right-censored survival analysis to avoid bias in your results.

Once you have the event and time-to-event columns, you have everything you need to perform a survival analysis. If you have data about a specific component and wish to know how the time to failure unfolds for each sample in the study, you can do a survival analysis and get a single Kaplan-Meier plot like *Figure* 4.19.

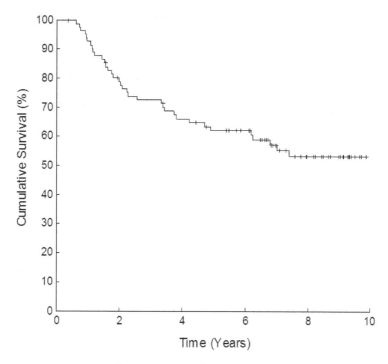

Figure 4.19: Kaplan-Meier plot

OK, if you've never seen a Kaplan-Meier plot before, the graph above might look a little strange.

Let's go through it step-by-step:

- The x-*axis* is the time difference from the first event (that is, the time of diagnosis) to the failure event (that is, death) or censor, and always begins at time t=0.

- The y-*axis* is the percentage (or proportion) of survivors remaining in the study at any given time point and the range is always [0, 100], or [0, 1] if shown as a proportion.

- At the beginning of the study (time = 0), there are no failure events.

- The graph always starts at the plot-point (0, 100).

- As time proceeds, the plot moves horizontally parallel to the x-*axis* until the next event (or censor).

- When an event occurs, the plot drops vertically in the graph by the proportion of the number remaining in the study at that point.

  For example, if at a given time-point, 1 patient out of 10 remaining in the study dies, the graph will drop by 10% from, say, 63% to 53%.

- When a censor occurs, a '+' is inserted onto the line to denote a censor, and the line continues horizontally.

The final point on the graph is always a '+' (the final data point is censored) or drops onto the x-axis (the final data point is a failure and there are no more patients remaining in the study).

The real power of this kind of analysis though is that you can use it to compare the survival rates of different groups or cohorts in the data. For this, you will need a third categorical variable, which can take two or more levels, such as Component [A; B].

The n×Categorical versus Survival

When you're running a survival analysis on a variable with n categories, what you will get is n survival curves, all on the same set of axes, as in *Figure* 4.20.

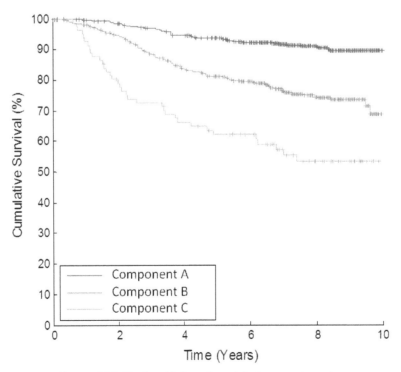

Figure 4.20: Kaplan-Meier plot with three categories

In this example, there are three groups of data corresponding to (supposedly) identical components manufactured by three different companies. The highest plot on the graph (Component A) corresponds to the highest probability of survival, whereas the lowest plot on the graph has the lowest probability of survival (Component C).

The kinds of questions you should be asking about these plots include:

- Is there a clear separation between the different groupings?

- Is the ordering of the groupings what you would expect?

- Do any of the plots cross over?

- Are there any large jumps in the plots towards the end (indicating a decrease in confidence)?

- Do any of the groupings drop to zero on the *y*-axis?

  Is this what you would expect?

- Are there any groupings that clearly clump together?

  Should they be aggregated; for example, to compare group 1 *versus* groups 2 and 3?

The data in *Figure* 4.20 is well behaved; there is a clear separation in the groupings, the separations are consistent, and there are no large jumps in the plots. Actually, this data isn't well behaved; it is *very well* behaved...

Perhaps the most appropriate statistic for this analysis is the Log-Rank Test, which tells us whether there is a significant difference between the plots.

The important outputs that you will need when you run a Kaplan-Meier survival analysis on n data categories are:

- **Visualization**: A Kaplan-Meier plot

- **Statistic**: A p-value

- **Effect Size**: The Hazard Ratio

The Kaplan-Meier plot is important because it gives you a good 'feel' for the data.

The p-value is important because it tells you whether there is evidence of a relationship between the variable and the probability of survival over a given time period.

The Hazard Ratio is important because it is the ratio of probabilities of failure. For example, a hazard ratio of two means that a group has twice the chance of dying than a comparison group. Although hazard ratios can only be calculated for two groups, it may be useful to compute the hazard ratio for each group compared to a baseline group. For example, if you designate group 1 as your baseline (most likely to survive), then you can compute the hazard ratio for groups 1 versus 2 and for groups 1 versus 3.

The analysis that accompanies Figure 4.20 is shown in *Figure 4.21.*

Censor Table:

	Total	Event	Censor	% Censor
Component A	580	52	528	91.03
Component B	449	107	342	76.17
Component C	82	35	47	57.32
Total	1111	194	917	82.54

Log-Rank Statistics:

χ^2	df	P
92.32	2	<0.001

Hazard Ratios:

	Hazard Ratio
Component A	1.00
Component B	2.92
Component C	6.20

Figure 4.21: Results of a Kaplan-Meier survival analysis

Looking at the % censor values of the three components, you can see that the chances of failure increase substantially as you go from Component A to B to C. This suggests that there may be a difference in survival between the three groups. You could do a Chi-Squared or Fisher's Exact test of the censor table and it would give you a useful snapshot of the study outcome at the final time-point, but it wouldn't give you any information about how the study events unfolded over time.

Running a Kaplan-Meier analysis tells you that for Components A and B, the likelihood of failure is approximately constant over the study period (the plots are roughly linear with no large jumps). For Component A, more than 90% of the components survived beyond 10 years, while around 75% of the components in category B survived to 10 years (you should ignore the steep jumps at the end of the plot – this is due to having few components left in the study and can bias the endpoint).

With Component C, roughly 30% of the parts failed within the first 2 years and 50% within the first 7 years. However, after 7 years, all of the components survived to 10 years.

The Kaplan-Meier plot and censor table suggest that there may well be a significant difference in survival between the groups, and the log-rank p-value statistic confirms this (< 0.001).

The hazard ratios tell you that, compared to Component A, Component B is almost three times as likely to fail (2.92), and Component C is more than six times as likely to fail (6.20).

The 2×Categorical versus Survival

If you have a variable with two categories, then what you'll get is a survival curve and associated stats for two groups rather than n groups.

However, a useful analysis is when you have n groups and you wish to make a more detailed comparison of one grouping of variables *versus* another grouping.

For example, there may be a clear distinction between a group of 'better surviving' components (perhaps denoted as 1-4) and 'worse surviving' (perhaps categories 5-10). In this case, you may wish to group together categories 1-4 into a single data category, do the same with categories 5-10, and run a Kaplan-Meier analysis with two categories of 'better components' and 'worse components.'

The advantage here is that you will get a clear result. The p-value and hazard ratio will clearly distinguish between the two groups and will give you a good understanding of the difference between the groups. Perhaps there are different processes underpinning the two different groupings.

This can set you off on an entirely new voyage of discovery...

The important outputs that you will need when you run a Kaplan-Meier survival analysis on two data categories are:

- **Visualization**: A Kaplan-Meier plot

- **Statistic**: The p-value

- **Effect Size**: The Hazard Ratio

The Kaplan-Meier plot is important because it gives you a good 'feel' for the data.

The p-value is important because it tells you whether there is evidence of a relationship between the variable and the probability of survival over a given time period.

The Hazard Ratio is important because it is the ratio of probabilities of death (or survival). For example, a hazard ratio of 3 means that a group has 3 times the chance of dying of the comparison group.

This concludes the chapter on univariate statistics. Next comes multivariate statistics. It's going to get a little bit harder, but wait, stop running – it's really not that difficult. Honest!

In *Chapter 5, Multivariate Statistics*, we'll go through the difference between univariate and multivariate stats, and you'll get an understanding of why you need more complex statistics to get a strong understanding of what's going on in your data.

5

Multivariate Statistics

In this chapter, you will learn that, while in univariate relationship analysis you are analyzing data pair-wise – in other words, *this* variable against *that* – in multivariate relationship analysis, you are assessing many variables (predictor variables) against a single variable (target, outcome, or hypothesis variable) simultaneously. In other words, you are testing *this*, *that*, and *the other* against a single outcome.

Testing multiple variables simultaneously has the distinct advantage that interactions between the variables can be controlled for.

OK, enough of the jargon – what does "controlled for" really mean?

Well, the bottom line is that univariate analyses do not take into account any factors other than the ones in the test. A univariate analysis of *this* against *that* tells you whether there is a relationship between a pair of variables, but it doesn't tell you whether that relationship is independent of other factors.

For example, if you find a significant relationship between *this* and *that* in univariate analysis, you can't be sure you're seeing the full story, because the influence of *the other* has not been tested for.

For this, you need to run a multivariate analysis, which distinguishes between those variables that are:

- not related
- dependently related; that is, the relationship depends on other factors
- independently related; that is, the relationship does not depend on other factors

Let's jump right in and have a look at the different types of multivariate analyses. After that, we'll deal with the difference between univariate and multivariate analysis and why you need a holistic strategy to discover all the independent relationships in your data.

Types of Multivariate Analysis

There are three types of multivariate analysis that you will use to find associations and correlations in your data:

- Logistic regression
- Multivariate linear regression
- Cox's proportional hazards survival analysis

Logistic regression is used when your outcome variable is categorical.

When your outcome has two categories, such as gender [male; female] or survival status [dead; alive], then you should use **binary logistic regression**.

If your outcome has more than two categories, you need to decide whether there is an order to the categories. If there is, such as when your outcome has categories of size [small; medium; large] or grade [A; B; C; D; E], then you should use **ordinal logistic regression**.

Alternatively, if your categories have no order, such as nationality [British, American, Spanish, ...] or religion [Hindu, Buddhist, Jedi], then you should use a **nominal logistic regression** (also known as multinomial logistic regression).

Multivariate linear regressions (also known as multiple linear regressions) are used when your outcome variable is continuous, such as height, weight, or some other such measurement.

When you have time-to-event data, such as that in a Kaplan-Meier analysis, then you should use **Cox's proportional hazards survival analysis**.

In each of these cases, you can use binary, ordinal, interval, and/or ratio data as predictor variables. Where you have nominal data, it may be useful to arrange it into separate variables, as in *Chapter 2, Data Classification*.

Binary Logistic Regression

Binary logistic regression is a special case of both ordinal and nominal logistic regressions. If your outcome variable has two categories, it doesn't matter whether you choose a binary, ordinal, or nominal logistic regression, the result will be just the same. In a sense, then, you might conclude that binary logistic regression is redundant, and you might be right, but nevertheless it's a good starting point.

Let's take a simple example to start us off on our multivariate journey.

Say you wish to find out whether smoking and/or weight are related to resting pulse. Your variables have the following properties:

- Outcome: Resting Pulse: binary, [high; low]
- Predictors: Smokes: binary, [yes; no]; Weight: continuous, measured in lb

Among the myriad of results that a typical commercial stats program will spit out at you (yes, I know, it can be very confusing and immensely frustrating), you should be able to find something that looks a bit like *Figure 5.1*.

Response Information:

Variable	Value	Count
Resting Pulse	Low (Reference)	70
	High	22
	Total	92

Binary Logistic Regression Table:

Predictor	Coef	SE Coef	Z	P	OR	L95	U95
Constant	-1.987	1.679	-1.18	0.237			
Smokes	-1.192	0.552	-2.16	0.031	0.30	0.10	0.90
Weight	0.025	0.012	2.04	0.041	1.03	1.00	1.05

Model Statistics:

G	df	P
7.574	2	0.023

Figure 5.1: Typical results of a binary logistic regression

The first thing to look at is whether your variables are significantly and independently related to the outcome. For this, you look to the p-values in the table.

In this case, both smoking and weight have p-values < 0.05, so they are both independently related to the outcome Resting Pulse.

The next thing is to notice the sign of the coefficient. The coefficient is a measure of effect size and the sign tells you whether the variable is positively or negatively associated with the outcome.

Also note that you use the coefficient to calculate the **Odds Ratio (OR)** with the following formula: $OR=e^{Coef}$.

Here, smoking is negatively associated with a low resting pulse. That's a bit awkward to wrap the brain cells around, so let's flip it around – smoking is positively associated with a high resting pulse. That makes more sense.

In terms of the OR, we can invert it, so 1/0.30 = 3.33. This means that smokers are more than three times as likely as non-smokers to have a higher resting pulse.

Similarly, weight is positively associated with a low resting pulse, but the coefficient is small (close to zero) and the OR is also small (close to one), so weight has only a small influence on resting pulse. In other words, a large change in weight is needed for a significant decrease in resting pulse.

The Constant is only important when creating a predictive model with your result (more on that in *Step-Wise Methods of Analysis* topic), and while you will use the coefficient, the p-value is not important and can be ignored.

One final point: the Standard Error of the Coefficient is used to calculate L95 and U95 – the Lower and Upper 95% Confidence Intervals (I'm leaving the equations out here – there are plenty of other texts that include them if you really need them). The confidence intervals give you the range of values that you can be 95% certain contains the true OR.

A tight range gives you high confidence, while a wide range gives you low confidence. A wide confidence interval is usually an indication that your sample size is too small for your analysis.

If your confidence intervals span the number one, then you cannot be certain that your variable will have any effect on the outcome and you should reject this result, even if it is statistically significant.

Looking at the model statistics below the table, you will see a separate p-value. This is the p-value for the whole analysis, and it tells you whether the model – in other words, the whole result – is statistically significant.

The model p-value is useful in comparing alternative models.

What happens to this p-value if you add or eliminate a variable? If the p-value becomes smaller, then you have a more accurate model.

The important outputs that you will need when you run a binary logistic regression analysis are:

- **Statistic**: individual p-values, model p-value
- **Effect Size**: coefficients, ORs, confidence intervals

The individual p-values are important because they tell you when your variables are significantly and independently related to the outcome.

The model p-value is important because it tells you whether the model is statistically significant and allows you to compare alternative models by the addition and elimination of variables.

The coefficients are important because they tell you how strongly the predictor variable is related to the outcome, and whether the relationship is positive (an increase in the predictor variable results in an increase in the outcome) or negative (an increase in the predictor variable results in a decrease in the outcome).

The OR is important because it tells you by how much the likelihood of the outcome changes with a one-unit change in the predictor variable.

The confidence intervals are important because they give you a sense of how confident you should be in the OR. If your 95% confidence intervals are very wide, you might wish to report 90% confidence intervals instead, or 99% confidence intervals when they are very narrow.

I mentioned earlier that a disadvantage of multivariate analyses is that they don't give you a good 'feel' for your data. Take another look at the results in *Figure 5.1*. Do you get a good sense of the mean value of weight? The maximum and minimum values? What about how many people smoke?

See what I mean? Multivariate analyses do have their advantages, but if you want to get a good feel for what's going on in your data, you'll need to do other stuff, such as descriptive statistics, graphs, tables, and univariate analysis.

Never jump straight into multivariate analysis without getting your hands dirty first, otherwise I guarantee you'll regret it (been there, got a truckload full of T-shirts!).

Ordinal Logistic Regression

Ordinal logistic regression is used when your outcome variable is categorical and has more than two levels, and there is some sort of natural order to the levels (but not necessarily the same distance between them).

Let's look at an example.

Say you wish to know whether there is an association between various factors on the survival of water lizards. You split your survival period into three levels 1, 2, and 3 corresponding to the short, medium, and long term, and a univariate analysis has revealed that there are two variables – toxicity and region – that may have an effect on survival. Your variables have the following properties:

- Outcome: Survival: ordinal, [1; 2; 3]

- Predictors: Region: binary, [1; 2]; Toxicity: continuous

Your results would look something like *Figure 5.2*.

Response Information:

Variable	Value	Count
Survival	1	15
	2	46
	3 (Reference)	12
	Total	73

Ordinal Logistic Regression Table:

Predictor	Coef	SE Coef	Z	P	OR	L95	U95
Constant (1)	7.244	1.879	-3.86	<0.001			
Constant (2)	-3.724	1.687	-2.21	0.027			
Region	0.201	0.496	0.41	0.685	1.22	0.46	3.23
Toxicity	0.121	0.034	3.56	<0.001	1.13	1.06	1.21

Model Statistics:

G	df	P
14.713	2	0.001

Figure 5.2: Typical results of an ordinal logistic regression

As before, look at the p-values of the predictor variables first. Toxicity is independently associated with survival but region is not. Region contributes only noise to the model, so you should now exclude it from the model and run it again. This would likely give you a more accurate model (see the *Step-Wise Methods of Analysis* section for more details on step-wise modeling techniques).

The toxicity coefficient is positive, telling you that there is a positive association between toxicity and survival; that higher toxicity levels in the environment tend to be associated with lower values of survival. Note that survival level 3 is designated as the reference, so the outcome survival is interpreted as the opposite of what is expected. If survival level 1 was taken as the reference, then the size of the coefficient would have been precisely the same but would have instead been negative (and you would have a reciprocal OR).

The OR tells you that there is a 13% increase in the water lizard survival odds for each one unit increase in toxicity; that is, a 13% increase in odds of survival for survival level 2 compared to 3 or level 1 compared to 2. Note that this is not the same as a 13% increase in the odds of absolute survival – you are not measuring absolute survival here, but rather categorizing survival as short, medium, or long.

No doubt you will have noticed that there are two constant coefficients here. These correspond to the coefficient for survival level 1 and survival level 2. The model could have included a coefficient for survival 3, but since this is the reference, its value would be 1.000, so there's no point including it.

Notice how similar the form of the ordinal logistic regression result is to that of binary logistic regression. It doesn't take a great leap of imagination to see that if you use ordinal logistic regression when your outcome variable has two categories, then the result will be precisely that of a binary logistic regression. This means that when your outcome variable is ordinal, then it doesn't matter how many outcome levels there are, the ordinal logistic regression is always applicable, even if there are just two outcome levels.

Nominal Logistic Regression

Nominal logistic regression is used when your outcome variable is categorical and has more than two levels, and there is no natural order to the levels.

Suppose that you want to know the favorite subject of a group of school kids, and your preliminary tests have identified that Age and Teaching Method are possible contributing factors in their choice of favorite subject. Your data has the following properties:

- Outcome: Subject: nominal, [science (reference); maths; art]

- Predictors: Teaching method: binary, [discussion; lecture], Age: continuous

Your results would look something like *Figure* 5.3 (for clarity and brevity, I've omitted the Response Information and Model Statistics from these results).

Nominal Logistic Regression Table

Logit 1: maths/science

Predictor	Coef	SE Coef	Z	P	OR	L95	U95
Constant	-1.123	4.564	-0.25	0.806			
Teaching (lecture)	-0.563	0.938	-0.6	0.548	0.57	0.09	3.58
Age	0.125	0.401	0.31	0.756	1.13	0.52	2.49

Nominal Logistic Regression Table

Logit 2: art/science

Predictor	Coef	SE Coef	Z	P	OR	L95	U95
Constant	-13.85	7.243	-1.91	0.056			
Teaching (lecture)	2.77	1.372	2.02	0.044	15.96	1.08	234.91
Age	1.014	0.584	1.73	0.083	2.76	0.88	8.66

Figure 5.3: Typical results of a nominal logistic regression

This time, there are two groups of results: a group comparing maths with science, and a group comparing art with science.

In the former group, according to the p-values, neither age nor teaching method contributed to the choice of students' favorite subject, and this is reflected in the OR 95% confidence intervals, which span unity.

However, when comparing science with art, the teaching method did make a significant contribution to their choice. The positive coefficient tells you that students given lectures rather than discussion-type teaching tended to prefer art to science – in fact, the OR suggests they are almost 16 times as likely to choose art over science when given lectures as opposed to discussion. Note though the wide span of the confidence intervals, suggesting that perhaps there isn't sufficient data to make any firm conclusions.

Also, take a look at the p-value for age. Although non-significant at the 95% level, given that there isn't much data available (there were only 30 students in the study), it would be reasonable to judge this data at the 90% level. This data suggests that a larger-scale study might provide additional insights.

An alternative way to have done these analyses would have been to create binary dummy variables for the nominal variable such that you would have a column for science [1; 0], one for art [1; 0], and one for maths [1; 0]. Then instead of a nominal logistic regression, you would run binary or ordinal logistic regressions. This would have the advantage that you would be comparing each of the nominal levels independently against all of the rest, rather than pair-wise as in the example of *Figure* 5.3. It all depends on what you're looking for. Do you want a pair-wise comparison, or one against the rest?

Multiple Linear Regression

Multiple linear regression is used when your outcome variable is continuous and has more than one predictor variable. Predictor variables can be continuous or categorical.

Let's say that you wish to assess the potential influence of various factors on systolic blood pressure. Your preliminary investigation has narrowed the possibilities to age and weight. Your variables have the following properties:

- Outcome: Systolic Blood Pressure: continuous, in mmHg

- Predictors: Age: continuous, in years, Weight: continuous, in pounds

Your results would look something like *Figure 5.4*.

Regression equation:

Systolic Blood Pressure = 30.990 + (0.861 x Age) + (0.335 x Weight)

Summary Table:

Predictor	Coef	SE Coef	T	P
Constant	30.990	11.94	2.59	0.032
Age	0.861	0.248	3.47	0.008
Weight	0.335	0.131	2.56	0.034

Regression Statistics:

S	R^2	R^2 (adj)
2.318	97.7%	97.1%

ANOVA:

	df	SS	MS	F	P
Factor	2	1813.92	906.96	168.76	<0.001
Error	8	42.99	5.37		
Total	10	1856.91			

Figure 5.4: Typical results of a multiple linear regression

The p-values tell you that both age and weight are significantly and independently associated with systolic blood pressure, and that systolic blood pressure increases with both age and weight (positive coefficients).

The adjusted R^2 value (which accounts for the number of predictors in the model) tells you that the predictors explain 97.1% of the variance in systolic blood pressure and only 2.9% remains unexplained – in other words, the model fits the data extremely well.

Note that there aren't any ORs in the table here. ORs are only appropriate when your outcome variable is categorical. Instead, when you want a measure of effect size, you look to the coefficients. In the *Correlations* topic, recall that one of the important measures in a univariate correlation test is the equation of the line. Well, its no different for the multivariate version. The equation of the line in a univariate linear regression takes this form: $y = mx+c$.

In a multivariate linear regression, there will be multiple independent predictors ($x1$, $x2$, $x3$, and so on) and coefficients (m_1, m_2, m_3, and so on), like this: $y = m_1 x_1 + m_2 x_2 + m_3 x_3 + ... + c$.

The coefficients are a measure of effect size and in this multiple linear regression example, they tell you by how much the systolic blood pressure increases for a single unit change in age or weight.

You will often find that an ANOVA has also been run on your data and the p-value you find there tells you how significant your model is, allowing you to compare other models by adding and eliminating variables.

Cox's Proportional Hazards Survival Analysis

Cox's proportional hazards survival analysis (also known as Cox's regression) is used when your outcome is a time-to-event variable.

For example, let's say you wish to find out what variables contribute to survival of colorectal cancer. Results of Kaplan-Meier survival analysis suggest that age, weight, and gender may each play a part, so you want to run a multivariate analysis. Cox's regression is the multivariate equivalent of the Kaplan-Meier analysis and is the appropriate choice. Your variables have the following properties:

- Outcome: Survival status: binary [dead; alive]; Time to event: continuous, measured in years

- Predictors: Age: continuous, measured in years; Weight: continuous, measured in kg; Gender: binary [male; female]

Your results would probably look a bit like *Figure* 5.5.

Response Information:

Variable	Value	Count
Death	1 (Event)	108
	0	349
	Total	457

Cox's Regression Table:

Predictor	Coef	SE Coef	Z	P	OR	L95	U95
Age	0.112	0.039	2.91	0.004	1.12	1.04	1.21
Weight	0.055	0.017	3.24	0.032	1.06	1.02	1.09
Gender (male)	0.074	0.188	0.39	0.694	1.08	0.75	1.56

Model Deviance:

G	df	P
12.57	3	0.006

Figure 5.5: Typical results of a Cox's regression

The p-values tell you that both age and weight are significantly and independently associated with death, and that the risk of death increases with both age and weight. The hazard ratios indicate that for each year older, the risk of death to the patient by colorectal cancer increases by 12%, and for each additional kilogram the patient weighs, the risk of death also increases by 6%.

Here, gender is not independently associated with death.

If you wish to accompany your Cox's regression results with Kaplan-Meier survival plots (and their hazard ratios), you may. It would be appropriate here to publish Kaplan-Meier survival plots of age and weight with your Cox's regression results, but not gender. As gender is not independently associated with survival, you can conclude that the relationship between gender and survival is dependent upon age and/or weight, so publishing a Kaplan-Meier plot of gender would be a misrepresentation of the data.

The strength of the overall model is known as the deviance and an overall p-value is calculated.

Using Multivariate Tests as Univariate Tests

You may already have figured this out by now, but you can use multivariate analyses as univariate analyses. All you need to do is put a single predictor variable in your model, and voilà – a univariate analysis.

The result you get will be the same as (or at least, very similar to) the result you would get from the equivalent univariate analysis in terms of the p-value, coefficients, OR, and so on.

Some researchers prefer to use multivariate analysis tools for their univariate analysis before going on to do a full multivariate analysis with the same tools. The advantage to this is that they get consistency in the tools they use.

I prefer not to do this, though. I *like* using different tools for univariate and multivariate analysis mostly because the univariate tools give you a better 'feel' for your data, but also because I feel more confident in the results if they concur and have been computed by different methods.

Univariate versus Multivariate Analyses

There is often a tension between the results of univariate and multivariate analysis that you need to understand a bit better, yielding questions such as:

- Why do the univariate and multivariate results differ?
- What does it mean when results are significant in univariate analysis, but not in multivariate analysis?
- More importantly, what does it mean when results are significant in multivariate analysis, but not in univariate analysis?
- What does it mean when there are two variables that are each independently associated with the outcome, but are also independently associated with each other?

There are basically three effects in the data that you need to know about to be able to explain why your univariate and multivariate results differ:

- Confounding variables
- Suppressor variables
- Interacting variables

Univariate and Multivariate Results that Concur

When the univariate and multivariate results concur, then we are living in happy times. Everything is just as it should be, there is no confusion in the results, and we can all ride off into the sunset knowing that we've done a good job.

How often does this happen? Almost never. This is why statisticians are often grumpy.

Confounding Variables

Confounding variables typically account for results that are significant in univariate analysis but are non-significant in multivariate analysis.

OK, let's cut through the jargon. What does confounding mean?

Well, the easiest way to explain is to look at an example. Let's have a look at a classic epidemiological example of confounding.

Let's say that in your analysis, you find that there is a significant relationship between people who have lung cancer and people carrying matches.

The obvious inference is that matches cause lung cancer, but we all know that this explanation is probably not true. So, we need to look deeper into the dataset to see what is responsible for this result, to find out which particular variable is likely to *confound* (cause surprise or confusion in) this result.

Digging deeper, we find that smoking is significantly associated with both lung cancer and carrying matches, as shown in *Figure 5.6*.

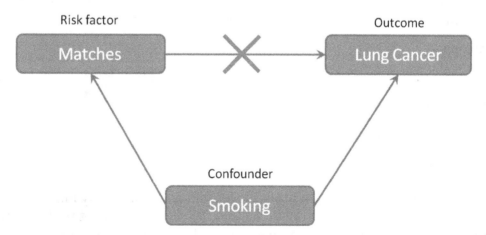

Figure 5.6: Illustration of confounding

The smoking variable now causes confusion in the relationship between matches and lung cancer, confounding our initial observations. The observation that there is a relationship between matches and lung cancer has become distorted because both are associated with smoking. In this example, smoking is called the confounding variable.

Confounding can really screw up your understanding of the underlying story of the data. Without an understanding of confounding variables in your analysis your results will be at best inaccurate and biased, but at worst completely incorrect.

Fortunately, there are a couple of analysis techniques that you can use to control for confounding:

- Stratification

- Multivariate analysis

Stratification allows you to test for relationships within the strata (layers) of the confounding variable.

For example, if you were to take your matches and lung cancer data and separate them into separate layers (data subsets), one for 'smoking' and the other for 'non-smoking', the association between matches and lung cancer can be tested within the smoking population and within the non-smoking population separately.

In this example, you will likely find that there is no relationship between matches and lung cancer in each of the strata, telling you that smoking is what is related to lung cancer and not matches.

Stratification has the drawback, though, of diminishing amounts of data as the depth of your strata increases. Imagine performing these analyses in the strata:

- Smoking

- Men that smoke

- Men over 50 that smoke

- Left-handed men over 50 that smoke

- Left-handed men over 50 with diabetes that smoke

To these factors, we can also add in fitness level, family history, body mass index, and a whole host of environmental and genetic factors. By the time you get to the bottom of these strata, there will be very few samples left, and in the time it's taken you to do the analyses, you will likely have grown old, grey, and very frustrated!

Multivariate analysis allows you to test for relationships while simultaneously assessing the impact of multiple variables on the outcome without having to limit the pool of data.

It tells you of the various risk factors and their relative contribution to outcome, and gets round the issue of confounding by adjusting for it.

Let's have a look at an example adapted from a paper by Hasdai, D., *et al* "*Effect of smoking status on the long-term outcome after successful coronary revascularization*". N. Engl. J. Med. 336 (1997): 755-761.

They ran their univariate analysis comparing the risk of death for four different cohorts – non-smokers, former smokers, recent quitters, and persistent smokers – and found that recent quitters and persistent smokers had a *decreased* risk of death compared to non-smokers.

What? How does this make sense? How can the risk of death be *lower* for persistent smokers than non-smokers?

To answer this question, they then ran multivariate analyses taking into account other factors such as age, diabetes, and hypertension.

The results comparing the univariate and multivariate analyses are shown in *Figure* 5.7.

		Non-Smokers	Former Smokers	Recent Quitters	Persistent Smokers
Relative Risk of Death	Univariate Results	1.0	1.08	0.56	0.74
	Multivariate Results	1.0	1.34	1.21	1.76

Figure 5.7: Comparison of univariate versus multivariate results

OK, the multivariate analyses make a lot more sense. All those that had smoked at some point in their lives had a higher risk of death than did non-smokers, and, just as we would expect, the persistent smokers had the highest risk of all.

So, what went wrong with the univariate analyses? Well, nothing really. It's just that the univariate analyses were *incomplete*. As it turned out, the recent quitters and persistent smokers were younger and had fewer underlying medical problems than the non-smokers had, and these factors confounded the univariate analyses, biasing the univariate results.

Hasdai and colleagues could have done a stratified analysis, but it would have been difficult to stratify for multiple variables. They were correct in choosing to run a multivariate analysis.

Suppressor Variables

When your results are significant in multivariate analysis but non-significant in univariate analysis, then you are most likely dealing with one or more suppressor variables – variables that increase the validity of another variable (or set of variables) by its inclusion in the analysis.

Let's have a look at another example. Suppose you wish to see whether height and weight are correlated with the race times of sprinters. If you find in a univariate analysis that weight is correlated with time, but height is not, you might be tempted to dismiss the influence of height on race times. The temptation would be to exclude height from the multivariate analysis and conclude that weight is the only variable that is independently correlated with time.

Now let's assume that height and weight are correlated with each other. Some of the variance that is shared between height and weight may be irrelevant to the outcome, time, increasing the noise in the analysis. Having both height and weight in the analysis allows the multivariate technique to identify the variance that is irrelevant, suppress the noise, and thereby lead to a more accurate analysis.

This is why it is called a suppressor variable, because its presence suppresses irrelevant variance on the outcome, ridding the analysis of unwanted noise.

Figure 5.8 shows an illustration of suppression.

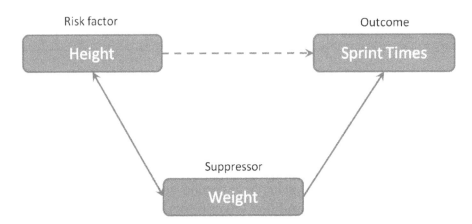

Figure 5.8: Illustration of suppression

Interacting Variables

In the case when you have two variables that are each independently associated with the outcome, but are also independently associated with each other, it is highly likely that these variables are interacting with each other in some way, modifying the effect of the risk on the outcome.

Let's go back to the confounding example, where the association between carrying matches and lung cancer was disproven by the confounding variable of 'smoking.' The link between matches and lung cancer was disproven because the association did not hold up in either of the smoking and non-smoking cohorts.

Now suppose that the association between matches and lung cancer holds up in both the smoking and non-smoking cohorts. This suggests that carrying matches and smoking each have a part to play in whether you get lung cancer.

If each of these variables plays a part, then there may be an additive effect. In other words, there would be a higher risk of contracting lung cancer if you carried matches *and* smoked, compared with carrying matches *or* smoked.

If you suspect that a pair of variables is interacting in some way, then you should test this out by creating an interaction variable.

The easiest way to do this is to create a single variable that represents the interaction. Code 'carrying matches *and* smokes' as 1, and code all other possibilities as 0.

Then add this into your multivariate analysis *in the presence of* both variables.

If there is an interaction between the variables, then the interaction variable will be statistically significant. Conversely, if the result is non-significant, then there is not likely to be an interaction between the variables.

For the results that are significant, if the variables act synergistically, then the correlation coefficient will be positive. Conversely, if the relationship between the variables is antagonistic (one works to decrease the effect of the other), then the correlation coefficient will be negative.

Figure 5.9 shows an illustration of interaction, where the outcome variable is modified (enhanced or diminished) by the presence of the interaction variable.

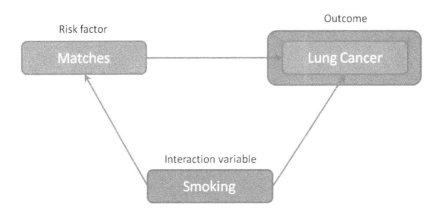

Figure 5.9: Illustration of interaction

If your variables interact, the nature of their interaction can be very complex, particularly if there are interactions between multiple variables, and it can be very difficult to figure out what is going on.

Dealing with interacting variables is best left to a specialist, and I'm not going to go into it in any greater detail here.

If you suspect that you have interacting variables, I would suggest you seek out an experienced statistician to help you. Go armed with a smile and a bottle of their favorite libation!

Limitations and Assumptions of Multivariate Analysis

Although each individual method of multivariate analysis has its own assumptions (discussed at the relevant point in the text), there is one assumption that is common to all, and that is the assumption of linearity.

The assumption is that the outcome changes linearly with each predictor variable. If the predictor variable is linear, then the assumption is that for a linear change in the predictor variable there will be a linear change in the outcome. When the predictor is ordinal, the size of the change in the outcome is the same for each unit change in the predictor.

What will likely change is the scale of the change, depending on the predictor variable. When variable A has a greater effect on the outcome than variable B, then a one-unit change in A will lead to a greater change in the outcome than a one-unit change in B. Each predictor variable may be weighted differently (the coefficients), but the assumption of linearity remains the same.

So, what happens when one of our predictor variables is not linearly related to the outcome?

Well, if you plot the predictor variable against the outcome, is the best-fit line *mostly* linear? If so, then it may be sensible to continue regardless.

On the other hand, if the best-fit line were clearly non-linear, then one way forward would be to transform either the predictor or outcome variable. If the outcome variable fails a standard normality test, then it may be sensible to transform the outcome; alternatively, you should consider transforming the predictor variable. Either way, you should repeat the univariate analysis once your variables have been transformed.

Typical transformations include:

- Logarithm (base 10)
- Natural logarithm (base e)
- Square root
- Square
- Reciprocal

There are many other more complicated transformations and I'm not going to delve deeper here.

Step-Wise Methods of Analysis

When you put all predictor variables into a multivariate analysis, what you'll typically get is a result where some of the variables are significantly associated with the outcome and some are not. Those that are not significantly associated with the outcome are adding noise into the model, which obscures the true signal. If you want to get a clearer picture, you want more of the signal and less of the noise.

If you identify the most non-significant predictor variable, exclude it and re-run the analysis you will most likely remove some of the noise, allowing the signal to come through more strongly.

Conversely, adding in an additional predictor that is significantly associated with the outcome will add to the signal and improve the signal-to-noise ratio in the model.

Step-wise methods of analysis are automatic procedures (although you can do them manually using most standard stats programs) that tell you which predictor variables should be in your multivariate analysis model.

These variable selection techniques have been much criticized as a modelling strategy, and rightly so – they have been much misused for decades. That doesn't mean, though, that you can't or shouldn't use them. There are various dangers in using step-wise procedures, but if you know them, you can account for them in your analysis to build a strong model and find all the predictor variables that are independently related to your hypothesis variable.

There are three ways to use step-wise procedures in multivariate analysis:

- Forward step-wise

- Backward step-wise

- Hybrid step-wise

Forward Step-Wise Procedure

With the forward step-wise technique, you start with no predictor variables in the model and add them one by one, like this:

1. Start with the univariate results and order the variables by p-value according to their relationship with the hypothesis variable.

2. Add in the predictor variable with the lowest p-value and run the model.

3. If the predictor variable is significantly associated with the outcome, leave it in the model for the next run, or kick it out.

4. Add in the predictor variable with the next lowest p-value and run the model.

5. If all predictor variables are significantly associated with the outcome, leave them in the model for the next run, or kick out the non-significant variables. You probably won't be able to predict which variables to kick out in advance because the multivariate model will take account of confounding and how closely related are the predictor variables with each other – the effects that univariate analysis didn't take account of!

6. Repeat *steps* 4-5 until there are no more predictor variables that remain un-analyzed.

Figure 5.10 may help you to understand the forward step-wise process.

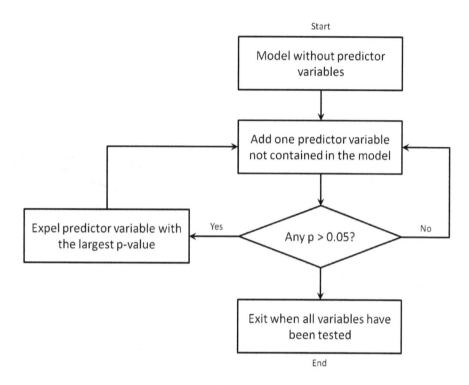

Start

Figure 5.10: Forward step-wise procedure

This procedure is good for when you have a small dataset, although in themselves small datasets may well give you spurious results in step-wise analysis. Tread carefully, folks!

A disadvantage is that suppressor variables might be eliminated from the model before they've had the chance to pair up with their partners in crime.

Backward Step-Wise Procedure

With this technique, you start with all your predictor variables in the model and kick them out one by one, like this:

1. Add in all the predictor variables. Typically, you would leave out any variables that were not significantly associated with your hypothesis variable from results of univariate analysis. If this makes you nervous (and it probably should), then you could include all those variables that had a p-value < 0.10. Run the model.

2. Eject the predictor variable with the highest non-significant p-value from the model and re-run the model.

3. Continue *Step* 2 until all non-significant predictor variables have been kicked out.

Figure 5.11 may help you to understand the backward step-wise process.

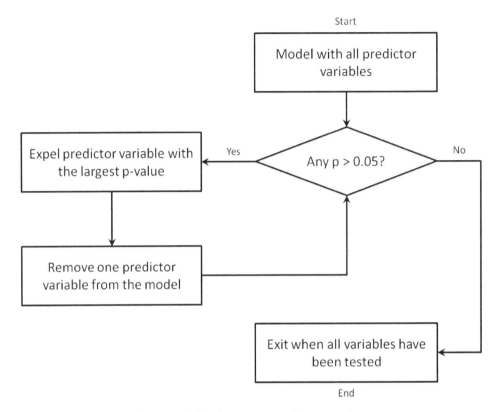

Figure 5.11: Backward step-wise procedure

This procedure gives suppressor variables the chance to do their thing, but has the significant drawback in that the early models containing all predictor variables will have a lot of noise in them and there is a real danger of eliminating an important variable from the model.

Hybrid Step-Wise Procedure

Hybrid procedures typically use a combination of backward and forward steps to try to get around the limitations of the backward and forward techniques.

If there are predictor variables that you know *should* be in the model, then you can usually tell the modeling software to start with those variables in the model and not eject them, even if they are non-significant.

Doing this, you can already see that you don't start with an 'all or none' model, but from somewhere in between; something like this:

1. Start with a core of user-selected predictor variables.

2. Add a forward step by adding a new predictor variable into the model.

3. Eliminate (backward step) until there are no non-significant variables in the model.

4. Add in a further new predictor variable and re-run.

5. Continue *steps* 2-4 until all predictor variables have been accounted for and there are only significant associations left in the model.

Figure 5.12 may help you to understand the hybrid step-wise process.

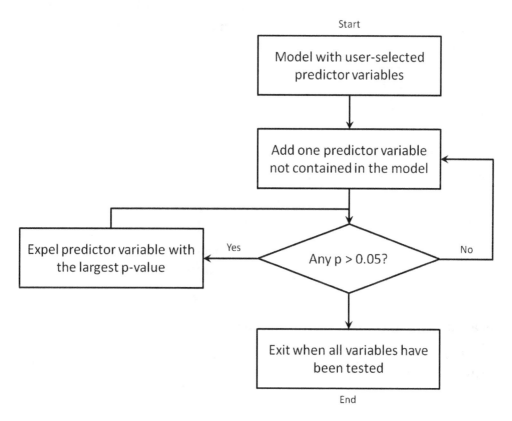

Figure 5.12: Hybrid step-wise procedure

There are as many variations of this as you can dream of, and many more additional checking steps you can add in. For example, once the model is complete, you might want to have another look at the non-significant predictor variables to see whether any of these might act as suppressor variables. Adding them sequentially to your final model to check their effect might just give you greater insights into their effect on your model.

Practical Tips for Step-Wise Procedures

If you read around the subject (yeah, I know, I really should get a life!), you'll see all sorts of rules of thumb to tell you how many samples you should have in a multivariate analysis compared to the number of predictor variables. The truth is that none of them are based in fact.

The bottom line is that it depends on the size of the effects you're measuring and how precisely you wish to measure them. Large effects can be detected with relatively small amounts of data, while small effects need large amounts of data. Some of your predictor variables might have large effects on the outcome, whereas others may simultaneously have small effects. It's complex and there is no one-size-fits-all solution to determining sample size for multivariate analysis.

If you suspect that your sample size is small compared to the number of predictor variables, one way to effectively handle your hybrid step-wise procedure is to separate your variables into pools, like you'll see at various sporting tournaments.

For example, let's say you have 85 variables:

1. You might choose to have no more than 10 predictor variables in any pool, so you'll have 10 variables in each of 8 pools and a ninth pool with 5 variables.

2. You can then run your favorite step-wise procedure on each pool until all non-significant variables have been eliminated from each pool.

3. The survivors go through to round 2, where new pools are created.

4. Repeat *steps* 1-3 until all predictor variables have been accounted for.

5. Go back through each eliminated variable testing to see whether adding it would improve the model.

Figure 5.13 may help you to understand the pooling process.

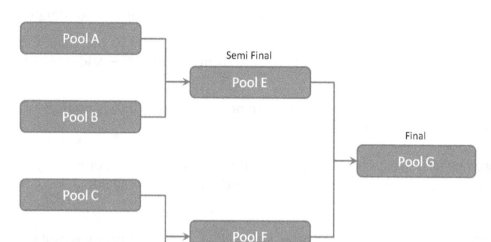

Figure 5.13: Pooling procedure

An issue that you'll come across when playing around with forward and backward elimination methods is that they don't always arrive at the same result. This can cause serious headaches and may well lead you to think that the method doesn't work and that it's more hassle than it's worth. I think it's fair to say that we've all thought that at some point. In fact, most of us have thought that about the whole of stats at some point in our learning curve!

When both your backward and forward analyses lead you to the same result, then you have earned yourself the rest of the day off. Relax, have a margarita, you deserve it – you have a robust model that you can take to the bank.

On the other hand, when they don't, then you still have work to do. Most likely, there will be suppressor effects or confounders that were present in the backward elimination technique but were kicked out of the forward procedure. Go and find them. Account for them.

There are many people who would advise you not to use step-wise procedures. I'm not one of those people.

My advice would be to learn how to use multivariate analysis techniques correctly and effectively – including the step-wise procedures.

There are dangers in using any tool incorrectly, and the best solution is to learn not walk away.

Using your knowledge and experience, address the following:

- Is your hybrid step-wise procedure not taking account of confounding variables? − Then re-model your procedure.

- Is a suppressor variable being eliminated from the model? − Add it to your core of non-eliminable variables.

- Are there interaction variables that are not being taken into account? − Create them and add them into the model.

There are no rules to say that the hybrid methodology should be completely automated, and nothing to stop you from adjusting the method and going again.

Experiment, learn, adjust, re-run. Rinse and repeat.

Whenever there is something itching at the back of your brain trying to tell you something is just not right, don't ignore it − scratch it! Find out what's going on and deal with it.

I guarantee that you'll learn a lot and end up with more accurate models than blindly accepting the first result − and that it's far better than walking away and not doing it at all!

Creating Predictive Models with the Results of Multivariate Analyses

So far, we've discussed the results of multivariate analysis in terms of correlations and associations, where the goal is to understand and explain the relationships. The next step is to use these results to create models that can predict the outcome variable given a new set of predictor measurements.

It may sound difficult, but it's really not. If you've run your multivariate analysis and got a good, sound result, then you have all you need to be able to make a good, sound prediction.

So far, so good, but notice that I said the result of finding the significant predictor variables in a multivariate analysis will be a *good, sound model*. I didn't say it would be the best or optimum model. Because the requirements of the multivariate technique are different when finding relationships to making predictions using them, your modeling method needs to be adjusted to account for that.

When seeking relationships, you look at the p-values of the individual predictor variables to try to find only those variables that are significantly and independently related to the outcome. In determining the optimum model, you're seeking to find the *optimum model*, not the constituent parts. To find the optimum model, you need to assess each forward or backward step in terms of the model p-value (or some other measure of model accuracy), not the individual p-values.

Whenever the addition or elimination of a predictor variable decreases the model p-value, the model has been improved and has greater predictive abilities than the previous model.

In most cases, eliminating non-significant predictor variables decreases the model p-value. However, there are times when the elimination or addition of variables has the opposite effect on the model p-value than would be expected by checking the individual p-value, and it is here that you need to remember the purpose of the analysis.

Predictive Models with Multiple Linear Regression

It all starts with the regression equation, which for a multiple linear regression looks like this: $y = m_1 x_1 + m_2 x_2 + m_3 x_3 + ... + c$.

We discussed this earlier: you take a new sample for which you have data on all of the predictor variables, say height, weight, age, and so on; multiply each of the predictor variables with its coefficient; add them up; and sum in the constant coefficient. This gives you a prediction for your outcome variable. If you already had a measurement for your outcome variable, but withheld it, you can compare the predicted outcome with the actual outcome to give you a measure of the error of the model.

Let's have another look at the example of **Systolic Blood Pressure (SBP)** in *Figure* 5.4. The regression equation for this was as follows: SBP=30.990+(0.861×Age)+(0.335×Weight).

The age was measured in years and weight in pounds, so a 50-year-old weighing 160lb is predicted to have an SBP of: SBP=30.990+(0.861×50)+(0.335×160)=127.6mmHg.

Ideally, SBP should be below 120mmHg, so according to the model, how can the patient reduce their SBP? Obviously, they cannot reduce their age, so they must lose weight. How much weight should they lose to lower their SBP to 120mmHg? Re-arranging the equation, the answer is found by this: Weight=((120-30.990-(0.861×50)))/0.335=137lb.

The patient must lower their weight to 137lb, which is a loss of 23lb. It looks like a more sensible diet and regular exercise is on the cards!

If, rather than requesting a prediction of the SBP, you wished to find the *change* in SBP between circumstances, you would use the following: Change(SBP)=0.861×(Age1-Age2)+0.335×(Weight1-Weight2).

This has the advantage that you can get a prediction for a change in outcome with a change in any predictor variable while keeping all other predictor variables constant. In this equation, this means that any variable that does not change drops out of the equation and simplifies the calculation.

Predictive Models with Logistic Regression

OK, what about logistic regression? Is that the same as multiple linear regression?

Well, nearly.

In logistic regression, the outcome variable is categorical, so instead we take the logarithm of the odds of having the outcome characteristic, which is also often called the logit of p in other textbooks. For example, say the outcome variable is lung cancer [yes; no]. The regression of the predictor variable will give us the natural logarithm of the odds of having lung cancer.

Remember that the odds are related to the proportion (or probability), p, of the outcome, y, by this: Odds=p/(1-p).

So, the form of a logistic regression is then as follows: loge (Odds(y))=m_1 x_1+m_2 x_2+m_3 x3+...+constant.

Let's have another look at the **Low Resting Pulse** (**LRP**) example of *Figure* 5.1. The regression equation of this model is then this: loge (Odds(LRP))=(-1.192×Smokes)+(0.025×Weight)+(-1.987).

For a non-smoker weighing 120lb, the regression equation becomes this: $\log_e$ (Odds(LRP))=(-1.192×0)+(0.025×120)+(-1.987)=1.013.

Taking the exponent of both sides of this equation gives this: Odds(LRP)=$e^{1.013}$=2.754 and Probability=Odds/(1+Odds)=2.754/(1+2.754)=0.734=73.4%.

So, someone who doesn't smoke and weighs 120lb has a 73% chance of having an LRP.

Now try plugging Smoking = 1 into the equation along with Weight = 120 and see what happens. Since the coefficient for smoking is negative, this should decrease their chances of having an LRP, and indeed that is exactly what happens; the probability drops to 45.5%.

Predictive Models with Cox's Regression

As with logistic regression, the outcome of the Cox's regression is a natural logarithm, but rather than being based on odds, it is now based on the ratio of the hazard (the probability of dying) $h(t)$ at time t to the baseline hazard, $h_0(t)$, like this: $\log_e (h(t)/(h_0(t)))=m_1x_1+m_2 x_2+m_3 x_3+...+m_kx_k$.

The baseline hazard is an abstract concept, and is the hazard when all the predictor variables are set to zero, such as when age = 0, weight = 0, and so on. Obviously, this is meaningless, so instead we can choose our own baseline levels and compare the hazard with these, like this: $\log_e (HR)=m_1 (xi_1-xj_1)+m_2 (xi_2-xj_2)+m_3 (xi_3-xj_3)+...+mk (xi_k-xj_k)$.

Wow, this is starting to look awkward. How did we get here? Well, what we did was take a hazard with a certain set of conditions, i, and set up the ratio of the hazard hi against the baseline hazard. We did the same again with a different set of conditions (j) and took the difference between them. This had the effect of removing the baseline hazard and giving us the **Hazard Ratio** (**HR**) between a pair of conditions that we specified.

It may look hard to work with, but don't worry, it's not as difficult as you think.

Let's go back to *Figure* 5.5 and take another look at the Cox's regression results.

If you took the baseline hazard to be a patient aged 50 with a weight of 60kg, what would be the likelihood of death for a patient aged 55 weighing 65kg? Plugging these values into the preceding equation would give you the following: $\log_e (HR)=0.112\times(55-50)+0.055\times(65-60)=0.837$ *and* $HR=e^{0.837}=2.30$.

This means that a patient aged 55 with a weight of 65kg is 2.3 times as likely to die as a 50-year-old patient weighing 60kg. Note that you don't say *when* the patient is likely to die. That's not what this means. The HR is the relative risk of a death occurring at any given time, and one of the assumptions is that anything that affects the hazard does so by the same ratio at all times (hence why it's called the proportional hazards model).

There is also another, simpler way of calculating this HR – by using the HRs reported in *Figure* 5.5. Since the per-year hazard of death is 1.12, the five-year hazard is calculated as 1.12^5. Similarly, the per-kg death hazard is 1.06, so the five-kg hazard is 1.06^5. Multiply these together and what do you get? HR = 2.30. Hazards are multiplicative!

This concludes your statistical toolbox. It's now time to put it all together, and this is exactly what we'll discover how to do in the next chapters.

Visualizing Your Relationships

In first two chapters of this book, you learned how to collect and clean your data, classify it, and prepare it for analysis, and then, you learned about the toolbox of statistical tests that you need to find the relationships that characterize the story of your dataset.

In this chapter, you will learn how to pull all of this together to create a holistic strategy for discovering the story of your data, using univariate and multivariate analysis techniques to discover which variables are independently related and using visualizations to interrogate these relationships in an intelligent, comprehensive, and inspiring way.

A Holistic Strategy to Discover Independent Relationships

So far, the tools that you have in your statistical toolbox have been seen as stand-alone, separate entities, and most statistics textbooks will present them to you in that way.

The truth, though, is that statisticians and experienced analysts use these tools in combination to give them a greater perspective on what the data is trying to tell them. And that is the crux of the matter here – data holds information hostage to the extent that it takes specialist tools and a joined-up strategy to extract the information and do something meaningful with it to change the world, hopefully for the better.

This is your mission, should you choose to accept it. This book will self-destruct in five seconds...

Univariate – Multivariate Tag Team

The typical reaction to looking at a whole dataset before starting any analysis is to start to have a panic attack. The whole thing is just too big to hold in your head at once, so you need to break it down into separate tasks, then into sub-tasks and sub-sub-tasks.

Let's start from the beginning, with the variable that you're most interested in. We'll call this variable A, or varA for short. You need to find out which variables are related to varA.

Way back in *Chapter 3, Introduction to Associations and Correlations*, we established that univariate statistics can tell you which of the other variables in your dataset are related to varA, but they do so pair-wise without correcting for the influence of other variables in the dataset. On the other hand, multivariate statistics can tell you which variables are independently related to varA – which is exactly what you want – but they are very data hungry and are sensitive to having too many variables in the model at once.

So, how can we find all the independent relationships with varA when there is not enough information going into univariate tests and too much going into multivariate stats?

You need a holistic strategy, and here it is:

- **Run univariate analyses of all the variables on varA**: Eliminate the variables with non-significant p-values.

- **Run multivariate analyses of the remaining variables with varA**: The significant relationships will be independently related to varA.

Essentially, you use univariate analyses as a screening tool to decide which of the relationships are *worthy of further investigation*. You're not making definitive decisions at this stage, but rather narrowing the field for multivariate analyses. You're using univariate and multivariate stats as a tag team; the univariates "soften up" the data, leaving them ready for multivariate stats to go in for the kill.

There are four different result possibilities when using univariate and multivariate statistics in this way, and here they are, with explanations of why you should accept or reject these results:

Non-Significance in Both Univariate and Multivariate Stats: Reject the Relationship

A relationship that is non-significant in both univariate and multivariate analysis is unlikely to be incorrect *in this dataset*, so you can safely reject this relationship as having no evidence to support its existence. That doesn't mean that the relationship doesn't exist in reality – your dataset may not have sufficient data to capture the complexities of the relationship, in which case you'll need more data or a bigger dataset.

Significance in Both Univariate and Multivariate Stats: Accept the Relationship

Similarly, a relationship that is significant in both univariate and multivariate analysis is also unlikely to be incorrect *in this dataset*, so you can safely accept this relationship as having some basis in reality. Again, that doesn't mean that the relationship is correct in the real world – it may have occurred by chance, and if you suspect that to be the case, then you'll need to run your project again with a new dataset to be able to confirm/deny this.

Significance in Univariate Stats and Non-Significance in Multivariate Stats: Reject the Relationship

What about when the relationship is significant in univariate analysis but non-significant in multivariate analysis? In this case, we can conclude that the relationship is not independent, but rather is dependent upon one or more confounding variables. Since you are only seeking independent relationships, you can reject this result.

Non-Significance in Univariate Stats & Significance in Multivariate Stats: Reject the Relationship

So, what do we do when the relationship is non-significant in univariate analysis but significant in multivariate analysis? In this case, most likely a suppressor variable is elevating the status of a relationship from being non-significant to significant. Although suppressor variables are important when trying to find the optimum *model*, here you are seeking only the independent relationships with varA. If varX is dependent upon a suppressor variable, varY, for its relationship with varA, then it is not independent and should be rejected. There is therefore no need to seek suppressor variables in the holistic relationship model.

However, if what you seek is the optimum predictive model of outcome variable varA, then suppressor variables are important. They must be identified and accounted for.

My advice here is to initially reject non-significant relationships at the univariate analysis stage and then do a post-hoc analysis after the multivariate analyses if you suspect that there are suppression effects. It can be fiendishly difficult trying to untangle the effects of suppressor variables, and if you're not comfortable (and I wouldn't be at all surprised), then seek out an experienced statistician to help you.

The bottom line here is that when seeking those variables that are independently related to your target variable, varA, you first reject all those variables that are not significantly related to varA in a univariate analysis. Of those that are significant in univariate analysis, you reject all those that are not related in a subsequent multivariate analysis. The remaining variables are those that are independently related to varA, which is precisely what you're looking for. *Figure 6.1* will help you decide whether to accept or reject relationships in the holistic relationship model.

		Univariate	
		Non-Significant	Significant
Multivariate	**Non-Significant**	Reject	Reject
	Significant	Reject	Accept

Figure 6.1: Hypothesis testing in the holistic relationship model

Tag Team Strategy for a Continuous Variable

Let's say your target variable, varA, is continuous. Right now, you have no idea which of the other variables in your dataset are related to varA. You'll probably have a hunch, and you'll have some prior knowledge from previous analysis, textbooks, and research publications, but these are all relationships that were found in *other* datasets. You don't know what you'll find in *your* dataset.

The first step in your holistic strategy is to use univariate analyses to find out which variables are related to varA. Remember that these won't necessarily be *independent* relationships, and that you're using univariate analysis as a screening tool to narrow the field to those variables that *might* be independently related to varA.

Taking note of the type of data for varA (in this case, continuous) and for varX, select the appropriate univariate statistical test. *Figure 6.2* will help you decide which test is appropriate when both your outcome and predictor variables are continuous.

		Outcome Variable	
		Continuous (Normal)	Continuous (Non-Normal)
Predictor Variable(s)	Continuous (Normal)	**Pearson Correlation** Scatter plot Correlation Coefficient Equation of the Line p-value	**Spearman Correlation** Scatter plot Correlation Coefficient Equation of the Line p-value
	Continuous (Non-Normal)	**Spearman Correlation** Scatter plot Correlation Coefficient Equation of the Line p-value	**Spearman Correlation** Scatter plot Correlation Coefficient Equation of the Line p-value

Figure 6.2: Appropriate analyses for a continuous outcome with continuous predictor

For example, if the predictor variable, varX, is continuous and both your outcome and predictor variables are normally distributed, you would choose the Pearson correlation.

On the other hand, if your outcome variable is continuous and your predictor variable is categorical, then *Figure* 6.3 will help you to decide correctly.

		Outcome Variable	
		Continuous (Normal)	Continuous (Non-Normal)
Predictor Variable(s)	Ordinal or Nominal (2 categories)	**2-sample t-test** Box-and-Whiskers plot Group means SD p-value	**Mann-Whitney U-test** Box-and-Whiskers plot Group medians Inter-quartile range p-value
	Ordinal or Nominal (>2 categories)	**ANOVA** Box-and-Whiskers plot Group means SD p-value	**Kruskall-Wallis** Box-and-Whiskers plot Group medians Inter-quartile range p-value

Figure 6.3: Appropriate analyses for a continuous outcome with categorical predictor

For example, if the predictor variable, varX, is ordinal and has two categories, you would choose the two-sample t-test or Mann-Whitney U-test, depending upon whether varA is normally or non-normally distributed.

After choosing correctly, run the test and inspect your p-value. If you want to be really strict, then you should reject those variables for which $p > 0.05$. These variables will not carry forward to a multivariate analysis. If you're a bit nervous about using such a strict p-value cut-off, then you can use one that is a little more lenient, such as $p > 0.10$. After all, the univariate tests are used as a screening tool here, and the final decision is made by the multivariate tests.

Continue to do this for varX1, varX2, varX3, and so on, until you have tested for all possible relationships with varA in univariate analysis.

You'll then have a list of possible relationships with varA that you'll need to test with multivariate analysis. *Figure 6.4* will help you decide which statistical test is appropriate.

		Outcome Variable	
		Continuous (Normal)	Continuous (Non-Normal)
Predictor Variables	**Multiple predictor variables**	**Multiple Linear Regression** Summary table Regression equation Regression statistics ANOVA	

Figure 6.4: Appropriate analyses for a continuous outcome with multiple predictors

Once you've used the multivariate tests on your data and you're completely happy with the results (check with an experienced statistician if you're not sure), then the relationships with varA could look something like *Figure 6.5*.

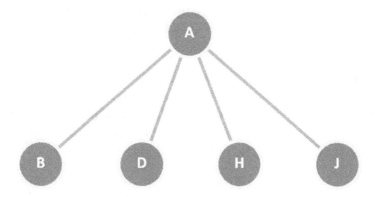

Figure 6.5: Possible relationships with varA after univariate analysis

Now that you have narrowed down the list of possible relationships, you can use multivariate statistics to confirm or deny these relationships. If your list of possible relationships is small, you can jump straight into multivariate analysis, otherwise you might need to split the variables into pools, as detailed in the *Practical Tips for Step-Wise Procedures* section of *Chapter* 5, *Multivariate Statistics*.

Using step-wise techniques, eliminate non-significant variables one-by-one until you are left with a core of variables that are all significantly and independently related with the outcome variable varA, say varB and varH, as in *Figure* 6.6.

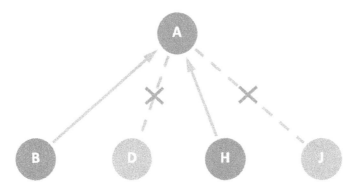

Figure 6.6: Relationships with varA after multivariate analysis

This is the story of the data for varA in this dataset. You have tested it with univariate analysis then confirmed it with multivariate analysis. You can be confident that these are all the independent relationships with varA in this dataset because you have accounted and corrected for all confounding variables.

Variables B and H are independently associated with A, while D and J are not – they are dependent upon other variables (B and H) for their relationship with A. Variables D and J are still related to varA and may still exert influence (and if you want the optimum predictive model, then you may need to have them in your final multivariate analysis), but their effects are secondary to the effects of B and H.

Note the directions of the arrows. One of the powerful aspects of multivariate analysis that you don't get with univariate analysis is that each of these variables are *predictors* of varA, that each of the relationships has a direction. You now know that each of B and H are independent predictors of A, but the opposite is not necessarily true – you haven't tested for the opposite yet.

Tag Team Strategy for a Categorical Variable

So, now you have the story of your data for variable A, let's look at the first of the variables that are independently related to it, varB, and let's suppose that varB has categorical data.

The strategy here is exactly the same as before; use univariate analyses to filter out the variables that are not related to varB, and then confirm/deny the relationships with multivariate analysis.

Figure 6.7 will help you decide which test is appropriate when your outcome variable is categorical and your predictor variable is continuous.

		Outcome Variable	
		Categorical (2 categories)	Categorical (>2 categories)
Predictor Variable(s)	Continuous (Normal)	**2-sample t-test** Box-and-Whiskers plot Group means SD p-value	**ANOVA** Box-and-Whiskers plot Group means SD p-value
	Continuous (Non-Normal)	**Mann-Whitney U-test** Box-and-Whiskers plot Group medians Inter-quartile range p-value	**Kruskall-Wallis** Box-and-Whiskers plot Group medians Inter-quartile range p-value

Figure 6.7: Appropriate analyses for a categorical outcome with a continuous predictor

Let's say that your outcome variable has two categories, such as Gender [Male; Female], and your predictor variable, varX, is continuous and non-normal. In this case, you would choose the Mann-Whitney U-test.

If both your outcome and predictor variables are categorical, then *Figure* 6.8 will help you to decide on the most appropriate statistical test.

		Outcome Variable	
		Categorical (2 categories)	Categorical (>2 categories)
Predictor Variable(s)	Ordinal or Nominal (2 categories)	**Fisher's Exact Test** 2x2 contingency table Chi-Squared cell values Odds Ratio p-value	**Chi-Squared Test** nx2 contingency table Chi-Squared cell values Chi-Squared Value p-value
	Ordinal or Nominal (>2 categories)	**Chi-Squared for Trend Test** 2xn contingency table Scatter plot of proportions Equation of the line p-value	**Chi-Squared Test** nx2 contingency table Chi-Squared cell values Chi-Squared Value p-value
	Nominal (>2 categories)	**Chi-Squared Test** nx2 contingency table Chi-Squared cell values Chi-Squared Value p-value	**Chi-Squared Test** nx2 contingency table Chi-Squared cell values Chi-Squared Value p-value

Figure 6.8: Appropriate analyses for a categorical outcome with categorical predictor

Run the appropriate relationship test and filter at your chosen p-value cut-off, and do this for varX1, varX2, and so on, until all possible relationships have been screened. Then you can run multivariate analyses on the variables that survive the cut. *Figure 6.9* will help you decide which statistical test is appropriate.

		Outcome Variable	
		Categorical (2 categories)	**Categorical (>2 categories)**
Predictor Variables	**Multiple predictor variables**	Ordinal or Nominal Logistic Regression Logistic regression table(s) Coefficients + SE Coefficients Odds Ratios + 95% confidence intervals p-values Model statistics	

Figure 6.9: Appropriate analyses for a categorical outcome with multiple predictors

The relationships with varB would then look like *Figure 6.10*.

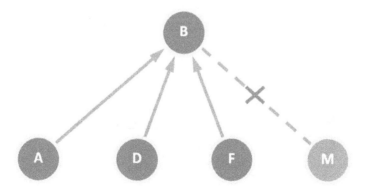

Figure 6.10: Relationships with varB after multivariate analysis

Figure 6.10 is the story of the data for varB, and all the relationships are significant and independent of other factors.

In this case, note that varA is an independent predictor of varB. In *Figure* 6.4, varB independently predicted varA, so the prediction was mutual. If you think about it long enough, you will realize that that doesn't have to be the case, though. Lung Cancer is a subset of Cancer, so Lung Cancer data will always predict Cancer (Lung Cancer 'Yes' always leads to a Cancer 'Yes'), but it's highly unlikely that the reverse will be true because there are many other subsets of Cancer (Cancer 'Yes' will lead to many more Lung Cancer 'No' than 'Yes'). It's a silly comparison to make because it wouldn't make sense to test for a relationship between Cancer and Lung Cancer, but I hope it illustrates the point that in relationship analysis, prediction is not necessarily mutual.

Tag Team Strategy for Survival Variables

For survival variables, the strategy remains the same. In this case, the univariate and multivariate tests are Kaplan Meier analysis (with the logrank test and hazard ratios) and Cox's regression, respectively. For Kaplan-Meier analysis, the predictor variable must be categorical. If you have a continuous predictor variable, then you must use Cox's regression, as either a univariate or multivariate test; see *Figure* 6.11.

			Outcome Variable
			Time-To-Event
Predictor Variable(s)	Univariate	**Ordinal or Nominal**	**Logrank Test** Kaplan-Meier Plot Censor Table Hazard Ratio(s) p-value
		Continuous	**Cox's Regression Analysis** Response Information Cox's Regression Table Model Deviance p-value
	Multivariate	**Multiple predictor variables**	**Cox's Regression Analysis** Response Information Cox's Regression Table Model Deviance p-value

Figure 6.11: Choosing appropriate analyses for a time-to-event variable

So, let's say that varC is your survival variable. As before, use univariate analyses (Kaplan-Meier for categorical predictor variables and Cox's regression for continuous predictor variables) to screen out the variables that are not related to varC, and then use multivariate analysis to confirm/deny whether the relationships are independent.

The relationships with varC might then look like *Figure 6.12*.

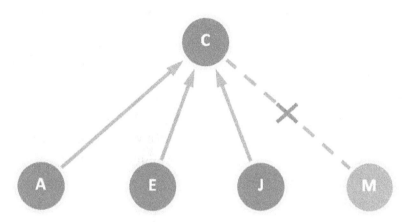

Figure 6.12: Relationships with varC after multivariate analysis

Here, varA, varE, and varJ are all independent predictors of the survival variable, varC, but varM is dependent upon at least one of A, E, and J for its relationship with C.

Corrections for Multiple Comparisons

It's round about this point that the more experienced analyst starts jumping up and down and getting all hot under the collar about multiple comparisons.

The crux of this argument is that if you run a single hypothesis test, there is a probability of 1 in 20 that a significant relationship will occur by chance. That is to say, that if you ran 20 hypothesis tests, then on average 1 will be incorrect. For 100 tests, 5 will be incorrect and for 1,000 tests, 50 will be incorrect. The argument goes that the more hypothesis tests you run on your data, the more false positives you will get in your result set.

I'm not convinced by these arguments. While the mathematics may appear sound, they tend to assume that chance is the most common explanation for the occurrence of a relationship, whereas the universe is governed by laws of nature and of physics. Most relationships are found because they have a basis in reality.

That said, let's look at the issue in practical terms. You compare varA with varB. Now compare varX with varY. Does the comparison between X and Y 'know' that the previous comparison has been made? Does it care? You cannot use quantum entanglement to explain how one hypothesis test might affect another.

OK, so you're not convinced yet.

Let's say you have a hypothesis about a particular relationship in a dataset, you do your analyses, adjust for multiple comparisons, and publish your paper. Later, you have another hypothesis that you can test within the same dataset. In this new set of analyses, should you adjust for the previous multiple comparisons you ran as well as the new ones? Now that you've run some new hypothesis tests, should you go back to your previous analysis and re-adjust according to the new tests you've just run? Should you adjust for *all* past tests that have been run on these data? If you adjust for all past tests, then surely you should adjust for all *future* tests, too! What if you have a dataset that is available to all researchers in your department – should you adjust for all past, present, and future tests made by your fellow researchers as well?

Starting to sound a little ridiculous, isn't it?

Well, let's go a bit further. Making adjustments for multiple comparisons can convince you to run an insufficient number of hypothesis tests on your data. If the relationship between varA and varB is borderline at $p = 0.046$, then any further tests will change your p-value from significant to non-significant, so you decide not to look for confounding variables or other potentially important relationships so you can preserve confidence in your result.

This really isn't a good idea. Why should the fear of losing a potentially important discovery stop you from being thorough in your analyses? It shouldn't. Worse still, it might lead to you not making other important discoveries in your data.

Perhaps the best way to account for false positive results is to use your domain knowledge and experience to ask of every relationship whether there is any reason to *reject* the result. Typically, whenever an unusual result occurs, the researcher will try to find a reason why it is plausible – and they will usually succeed. After all, we all love a winner, don't we? And if she's right, they can publish this new finding and enhance their career, gain new funding, and climb onward and upward.

It is better, though, to do the opposite and try to find – for each result – a reason why it is implausible. This way, you are building strong controls into your analyses and are making reasoned judgements that give a stronger basis for your results than arbitrarily adjusting p-values.

If you still believe in the concept of adjusting for multiple testing (I remain far from convinced), the question you must ask yourself is this: is it better to miss nothing real than to control the number of false positive results, or control the probability of making a false claim?

It is worth noting that most spurious results are accounted for by multivariate analysis, and if you insist on making adjustments for multiple comparisons, then I suggest you keep it simple and apply a Bonferroni correction.

Basically, a Bonferroni correction penalizes you for the number of pair-wise comparisons by requiring that you multiply your resultant p-value by the number of comparisons made. If, after making this adjustment, your p-value remains <0.05, then you may reject the null hypothesis and accept that the relationship is significant.

In the holistic strategy outlined here, I recommend that you don't apply any corrective factors because of univariate analyses, since the multivariate analyses will naturally correct for these.

You may apply corrective factors to your multivariate analyses if you wish, and I recommend that you correct for the number of independent predictors remaining in the final model.

For example, for varA, if there are five independent predictors after adjusting for all other factors, then multiply each of the p-values by five. The variables that have adjusted p-values larger than 0.05 will then be rejected. Moving on to varB, if the final model has three predictors, then multiply each of the p-values by three, and so on. Here, it doesn't make sense to penalize the results of varB because of the results of varA, so I see no reason to combine the penalties.

Visualizing the Story of Your Data

When you have many relationships between the variables in your dataset, it can be difficult, if not impossible, to hold it in your mind. It's not usually difficult to visualize the story of the data for varA; after all, just how many causal (or rather *perceived* causal) connections can there be for a single variable? However, when you pull the whole thing together, it can rapidly get away from you. Here's where you need to be able to build a visual representation of all the relationships in the dataset, and I'll present a couple of possibilities to you here.

Node-Link Diagrams

So far, you've investigated the relationships with varA and you have the story of the data for varA (*Figure* 6.4). Similarly, you've done the same for varB (*Figure* 6.10), varC (*Figure* 6.12), and so on, and it's now time to pull the whole thing together; see *Figure* 6.13.

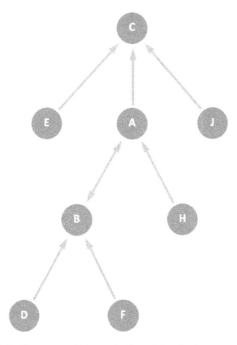

Figure 6.13: Node-link diagram of the relationships between varA, varB, and varC

In *Figure* 6.13, I've shown only the results for varA, varB, and varC, but, of course, the bigger picture will be much larger than this – we haven't yet considered the results of varD, varE, varF, and so on, but you get the idea. You take the story of A and make the connections where necessary to the other individual stories until you have the story of the whole dataset.

There are some amazing free visualization programs that will do this for you once you input your results. Typically, all you need to input is a list of the variable names and the relationships and the software does the rest. There are also many customization options in node-link diagram programs, such as:

- Adjust the thickness of links to signify the strength of the relationship (effect size)
- Adjust the size of the nodes to signify the importance in the network
- Adjust the color of nodes to signify certain data groupings

Node-link diagrams are a great non-linear structure of the visualization of the relationships between variables, and the available software programs often allow force-directed algorithms to spatially optimize the nodes and links in an aesthetically pleasing conformation. The purpose of these algorithms is to position the nodes in space so that all the links (relationships) are of more or less equal length and there are as few crossing lines as possible. By assigning forces among the set of links and the set of nodes, based on their relative positions, the forces can be used to minimize the energy of the arrangement.

With node-link diagrams, it is straightforward to zoom in to inspect detail, zoom out to see the bigger picture, select individual nodes to visualize only their first and second-level connections, or select any pair of nodes to inspect the possible pathways between them.

However, as networks get large and highly connected, node-link diagrams often devolve into giant hairballs of line crossings and can become practically impossible to work with.

Correlation Matrix

With node-link diagrams, only the significant independent relationships are displayed. If you want to visualize *all* the relationships in your dataset, whether they are significant or not and independent or not, the correlation matrix is probably the best way to go.

The correlation matrix is a square array with the variables placed along horizontal and vertical axes, like that shown in *Figure 6.14*.

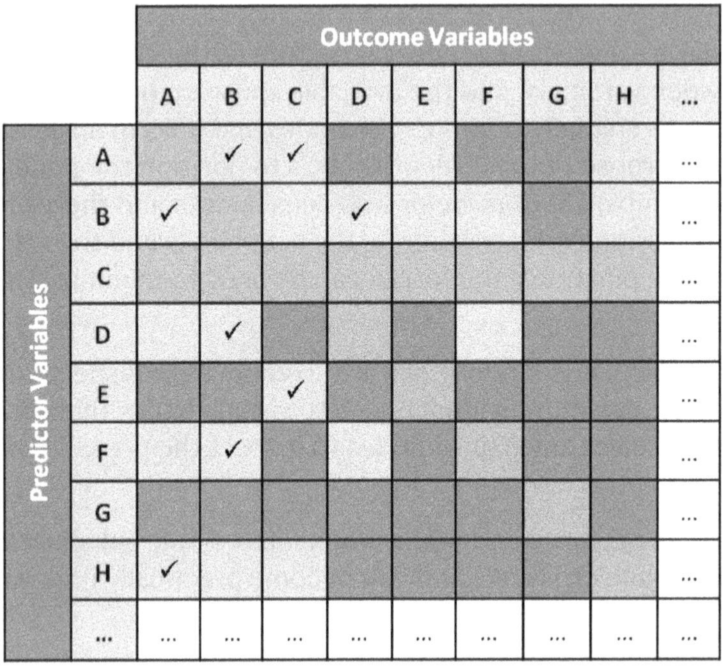

Figure 6.14: Correlation matrix of the relationships in the dataset

The outcome variables are on the horizontal axis, while the predictor variables are on the vertical axis. Each column is a list of all the relationships between the outcome variable and the predictor variables as determined by the results of the multivariate analysis, and each cell of the matrix represents the relationship status between the pair of variables. In this example, the status of the relationship is represented by traffic light colors; green is significant ($p < 0.05$, and marked with a ✓ to aid those with red-green color blindness), red is non-significant ($p \geq 0.10$), and amber tells you that the result is marginal and is perhaps worthy of further investigation ($0.10 < p \leq 0.05$). These cut-off values can be adjusted to whatever suits your purpose and the color scheme can be varied.

For example, we established in *Figure 6.12* that varA, varE, and varJ are independent predictors of varC. This is reflected in the column of varC: varA and varE are colored in green (and varJ would have been too if I'd made *Figure 6.14* bigger!). On the other hand, since varC is time-to-event data we can only analyze this data as an outcome, not as a predictor, so the entire row for varC is grayed out, signifying that no analyses have been performed or that analyses were not possible .

Correlation matrices are very informative because you can instantly see which of the variables in the dataset are important and which are not; for example, varG has no significant relationships with any other variables in the dataset.

Although for simplicity I've shown the cells here as being colors representing relationship status, you could also add into the cells the p-values giving rise to these relationships or perhaps odds ratios to signify the strength of the relationship.

Correlation matrices have an advantage over node-link diagrams in that line crossings are impossible with matrix views, although path following is harder in a correlation matrix than in a node-link diagram.

Diagrams such as these can be very important because they can tell you which other variables might be confounders (although the multivariate analyses should have accounted for these), or involved in suppression or interaction effects – they're not likely to be more than one degree of separation away from the outcome variable.

They're also important to give you a greater perspective of the story of the data. So often, researchers worry when a predictor variable that they expect to be independently related to the outcome is expelled from a multivariate analysis. They think it's a crisis. Being expelled from a multivariate analysis is not the end of the world and it doesn't mean that the predictor variable is having no effect. Showing how all the relationships interact with each other in a node-link diagram puts all the variables in their place, and should put you more at ease with the results.

Bonus: Automating Associations and Correlations

In this chapter, you'll learn why it's becoming increasingly difficult to do relationship analyses manually and why automation of the whole process is important, and be introduced to a product for just such a task – CorrelViz.

Disclaimer – CorrelViz is our own product and is available exclusively via Chi-Squared Innovations.

What is the Problem?

There is no shortage of large, do-it-all statistics programs that contain all the statistical tests you need to do your associations and correlations (and 10,000+ other statistical tests all rolled up into one huge package).

The problem with these is that you have to run the analyses manually. OK, sure, it's not terribly difficult to run a two-sample t-test or an ANOVA – they're just a few clicks away – but let's have a look at this in the context of the whole dataset.

Let's take a dataset of 100 variables (columns). That's pretty small by today's standards. The number of possible relationships in this dataset is (100–1) + (100–2) + (100–3) + ... + (100–98) + (100–99) = 4,950.

Now let's suppose that it takes an average of just five mouse clicks to locate and select your chosen test, select your variables, and click GO. That means that the story of your data is 4,950 x 5 = 24,750 mouse clicks away. *Gulp.*

But what if you then want to find whether the same relationships exist only in the male cohort? That's another 4,851 possible relationships and another 24,255 clicks. Now in just the female cohort? Another 4,851 (24,255 clicks). What about in the smoking cohort, the non-smoking cohort, the diabetic cohort, and the non-diabetic cohort? That's another 4 x 4,851 x 5 = 97,020 mouse clicks. You already have 34,056 possible relationships (with 170,280 mouse clicks) and you've barely scratched the surface.

And to think you started out with just 100 variables!

How long do you think it's going to take you to make nearly a quarter of a million clicks?

Even though most of us will never have to deal with big data, we all have small data, and small data is getting bigger and it's coming at us faster. Only a decade ago, analysts would often be heard exclaiming, "you've got data!," and now they're all terribly over-worked because everybody has data and they all want it analyzed yesterday.

So, if you have data and the commercial stats programs are too slow, how do you complete your analyses before dying of old age? Until recently, the answer has been that you compare a few variables with your hypothesis variable, check out a few inconsistencies, run a couple of subset analyses, and then write your paper or thesis. Basically, the researcher cherry-picks the analyses that they want to do, because they have neither the time nor the will to do a complete analysis. And it'll still take months to get all the analyses done!

In my experience, less than 10% of any dataset is adequately analyzed, and less than 1% is published! That's an enormous amount of information that we're missing.

Just imagine the discoveries that we could make with the data that already exists.

Just imagine the discoveries that you might be missing in *your* dataset.

CorrelViz

In the data-rich world that we're living in, the old ways of doing analyses are increasingly letting us down, and you need a faster, more efficient way of finding answers to your questions.

Most of the commercial statistical programmes allow you to program your own macros, and there are whole programming languages dedicated to allowing you to create your own statistical routines. That's great, but most people that have to do statistics can't program, have no incentive to do so, and certainly don't have the time.

How many surgeons do you know that can program in R or Python? How many microbiologists can program in SPSS or SAS? I've never met any.

There's never been a more pressing need for a fully automated dedicated solution to the problem of relationship analysis than right now.

Fortunately, there is a solution.

CorrelViz is a fully automated statistical analysis and visualization program that cleans and classifies your data, screens potential relationships by using the appropriate univariate statistical tests, confirms or denies these relationships with multivariate stats, and presents you with an intuitive, interactive visualization of the story of your data.

All this is completely automatic and takes you from data to story in minutes, not months.

Best of all, CorrelViz flips the methodology of data analysis on its head. Instead of doing your analysis first and then trying to visualize your results last, in CorrelViz, once your data has been cleaned and classified, all the data analysis runs automatically and the first thing you see is a node-link diagram of all the independent relationships in the dataset. You get the story *first*!

The visualization is interactive, so clicking on a link will give you the statistical details of the independent relationship between the pair of variables. Clicking on a node gives you all the details of which predictor variables are independently associated with your chosen outcome variable.

Through all this, there is no manipulation of data, no selecting incorrect statistical tests, no worries about confounding variables, and the whole process takes just minutes. Extract the relevant results and your whole research effort could be over before lunch!

Try it out here: https://chi2innovations.com/correlviz/.

Appendix

This section is included to assist the learners to perform the activities present in the book. It includes detailed steps that are to be performed by the learners to complete and achieve the objectives of the book.

Summary: A Holistic Strategy to Discover the Story of Your Data

Well, here we are at the end. I hope you've found something of value in this book.

Before we finish, let's just recap the steps taken to discover the story of your data:

1. Generate a hypothesis and start to *think* about the data that you'll need to collect in pursuit of the answer(s) that you seek. Get things wrong at this stage and the rest of your study could be useless. You don't want to go to your statistician after 3 years of study, only to be told that your dataset is not fit for purpose and cannot answer your hypothesis.

2. Collect your data *sensibly*, considering which relationships in your data would be worthwhile testing and which would be nonsense. Consider excluding the variables that defy logic.

3. Clean your data, correcting entry errors and illogical data such as patients that are -12 or 323 years old. Make sure that your dataset is in the correct format and amenable to enter into your chosen analytics package.

4. Classify your data, identifying those variables that are ratio, interval, ordinal, and nominal. The way you identify your data determines your choice of statistical tests, so it's crucial to understand your data.

5. Understand the purpose of your analysis. If you are seeking independent relationships with your hypothesis variable, you will create a different analytical strategy to that of seeking the optimum predictive model of your hypothesis variable.

6. Create your holistic analytical strategy in pursuit of your hypothesis. Your analysis will likely take many months, so write it down so that you understand what you're doing and check it with a statistician before you start. Your strategy will likely involve using univariate statistics as a screening tool and multivariate statistics to confirm or deny the independent relationships. If you are seeking the optimum model, you may need to do a post hoc analysis to account for suppressor and/or interaction effects.

7. Start analyzing. First, use univariate statistics to weed out those relationships that are not related to the outcome variable. The survivors will pass to a multivariate test to determine whether the relationships are independent (significant) or dependent upon other variables (not significant).

8. Perform any necessary post hoc analyses.

9. Create a visualization of the story of your dataset.

10. Shortcut *steps* 3-9 by using CorrelViz!

11. Publish your paper in the leading journal of your field and be the envy of your colleagues!

Epilogue

Final point: try to have fun with your data and analyses. Statistics is not a closed shop and not just for statisticians. Ask "what if..." of your data and play around with analyses in your pursuit of knowledge and understanding.

If you have categorical data that has, say, more than a dozen or so categories, what happens when you treat it as continuous data and analyze it as such? Try it. See what you can learn. What happens when you have an ordinal outcome variable and analyze it using a nominal logistic regression? Give it a go. You might learn something new.

Don't be afraid of trying something out. If you've never done a data transformation before but feel that your analysis might benefit from it, then give it a go. Play around with a few different transformations to get a *feel* for things.

Discuss your ideas and experiments with a statistician. Build your understanding. Be a better analyst. Most of all, don't be afraid to make mistakes – you will learn more from your failures than from your successes. I guarantee you'll learn a lot and you might just enjoy data analysis more than just turning the wheel with your nose to the grindstone.

If you've enjoyed this book, please feel free to contact me and tell me so. Conversely, if you haven't, then I would *still* love to hear from you (but be nice, please!): ebookfeedback@chi2innovations.com.

Copyright

Index

About

All major keywords used in this book are captured alphabetically in this section. Each one is accompanied by the page number of where they appear.

www.ingramcontent.com/pod-product-compliance
Lightning Source LLC
Chambersburg PA
CBHW080537060326
40690CB00022B/5155